PERSPECTIVES IN QUANTUM CHEMISTRY

ACADÉMIE INTERNATIONALE
DES SCIENCES MOLÉCULAIRES QUANTIQUES

INTERNATIONAL ACADEMY
OF QUANTUM MOLECULAR SCIENCE

PERSPECTIVES IN QUANTUM CHEMISTRY

PLENARY LECTURES PRESENTED AT THE
SIXTH INTERNATIONAL CONGRESS ON QUANTUM CHEMISTRY
HELD IN JERUSALEM, ISRAEL, AUGUST 22–25, 1988

Edited by

JOSHUA JORTNER
School of Chemistry, Tel Aviv University, Tel Aviv, Israel

and

BERNARD PULLMAN
Institut de Biologie Physico-Chimique
(Fondation Edmond de Rothschild), Paris, France

KLUWER ACADEMIC PUBLISHERS
DORDRECHT / BOSTON / LONDON

Library of Congress Cataloging in Publication Data

International Congress on Quantum Chemistry (6th : 1988 : Jerusalem)
Perspectives in quantum chemistry : proceedings of the plenary lectures at the Sixth International Congress on Quantum Chemistry held in Jerusalem, Israel, August 22-25, 1988 / edited by Joshua Jortner and Bernard Pullman.
p. cm.
Includes index.
ISBN 0-7923-0228-1
1. Quantum chemistry--Congresses. I. Jortner, Joshua. II. Pullman, Bernard, 1919- . III. Title.
QD462.A1I57 1988
540.2'8--dc19 89-2398

ISBN 0-7923-0228-1

Published by Kluwer Academic Publishers,
P.O. Box 17, 3300 AA Dordrecht, The Netherlands.

Kluwer Academic Publishers incorporates
the publishing programmes of
D. Reidel, Martinus Nijhoff, Dr W. Junk and MTP Press.

Sold and distributed in the U.S.A. and Canada
by Kluwer Academic Publishers,
101 Philip Drive, Norwell, MA 02061, U.S.A.

In all other countries, sold and distributed
by Kluwer Academic Publishers Group,
P.O. Box 322, 3300 AH Dordrecht, The Netherlands.

Printed in The Netherlands

TABLE OF CONTENTS

PREFACE

The Sixth International Congress on Quantum Chemistry convened at the Campus of the Hebrew University, Jerusalem, Israel, on August 22-25, 1988. The International Congresses on Quantum Chemistry are held under the auspices of the International Academy of Quantum Molecular Science. Previous International Congresses on Quantum Chemistry were held in France, Japan, the United States, Sweden and Canada. These prestigious meetings provided a central contribution to the important modern area of theoretical chemistry.

The major goals of the Sixth International Congress on Quantum Chemistry were:

A) To provide an overview of recent novel developments, advances and directions of research in the broad area of quantum molecular sciences.

B) To establish strong interaction between the theoretical discipline of quantum molecular sciences and experiment.

The general topics of the Sixth International Congress were:

a) Molecular Quantum Mechanics
b) Many-Body Theory of Molecular Structure
c) Intermolecular Forces
d) Complexes and Clusters
e) Molecular Spectroscopy
f) Intramolecular Dynamics
g) Chemical Reactions
h) Molecular Dynamics Simulations
i) Condensed-Phase Chemistry
j) Surface Phenomena and Catalysis
k) Quantum Biochemistry
l) Biophysics

The format of the Sixth International Congress consisted of plenary lectures, symposia and poster sessions. In the opening session of the Congress, commemorative addresses were delivered in honoured memory of the late Louis de Broglie and the late Robert S. Mulliken, Nobel Prize Laureates and Members of the International Academy of Quantum Molecular Science. A commemorative symposium was devoted to the honoured memory of the late Massimo Simonetta, Member of the International Academy of Quantum Molecular Science. The closing session of the Congress was dedicated to a symposium on the topic "Sixty Years Since the Heitler-London Paper", commemorating this seminal event in the development of modern quantum chemistry.

The modern trends and developments in theoretical chemistry were deliberated by intensive and extensive interaction between international scientists from many diverse areas between theory and experiment. This volume provides a record of the plenary lectures at the Sixth International Congress on Quantum Chemistry.

Held under the auspices of the International Academy of Quantum Molecular Science and the Israel Academy of Sciences and Humanities, the Sixth International Congress on Quantum Chemistry was sponsored by the Fondation Edmond de Rothschild and supported by the Israeli Ministry of Science and by the Israeli Universities. We are indebted to the Hebrew University, the Technion - Israel Institute of Technology, the Weizmann Institute of Science, Tel Aviv University and Bar-Ilan University for their support. We would like to convey special thanks and appreciation to the Hebrew University for hosting the Congress. Thanks are due to the Israeli Ministry of Science and the National Research Council for their support. The Sixth International Congress on Quantum Chemistry was most generously supported by the Fondation Edmond de Rothschild and we are greatly indebted to Baron Edmond de Rothschild for his remarkable support of this important scientific endeavour. We are also greatly indebted to Baron Edmond de Rothschild for his magnificent involvement in this Congress, which will contribute significantly to science.

Joshua Jortner
Bernard Pullman

Jerusalem and Paris
December 1, 1988

OPENING ADDRESS

JOSHUA JORTNER

The Israel Academy of Sciences and Humanities

On behalf of the Israel Academy of Sciences and Humanities, it is a great privilege and pleasure to extend a warm welcome to the Distinguished President and Members of the International Academy of Quantum Molecular Science, and to the Distinguished Guests and Participants at the Sixth International Congress on Quantum Chemistry (6th ICQC). We extend our most cordial greetings to all the scientists here on the Campus of the Hebrew University, who have gathered from far and near to attend the 6th ICQC. It is an honour for the Israel Academy of Sciences that this notable Congress is held under its auspices. The cultivation, enhancement and development of scientific values is truly international and universal, and there are no boundaries to intellectual, scholarly and scientific accomplishments. In this spirit of international scientific collaboration, I would like to thank all the visiting scientists and the Israeli scientists attending the 6th ICQC. Scientific and scholarly co-operation between scientists of various countries constitutes a major component of international relations. This is evident not only in the quantity but, first and foremost, in the quality of international scientific interaction collaboration. This Congress contributes significantly to high quality international scientific collaboration.

This central scientific endeavour constitutes an important event for the international scientific community of theoretical chemistry and chemical physics, as well as for the Israeli scientific community. The privilege to host this International Congress marks the strong commitment of the State of Israel and of our scientific community for the advancement of scholarly and scientific values, which is in line with the long cultural tradition of the Jewish people. All of us have to be committed to the advancement of scientific research for the acquisition of knowledge, which constitutes a noble goal of man, as well as for the utilization of wise application of research for the alleviation of the conditions of mankind.

It may be appropriate on this auspicious occasion to dwell on some reflections on the development of theoretical chemistry in Israel. When the Department of Physical Chemistry was established at the Hebrew University in 1934 by Ladisleous Farkas, who was the discoverer of the dynamic effects of predissociation, a substantial amount of the scientific research effort was directed towards chemical physics, pertaining to the problems such as spectroscopic and thermodynamic properties of orto- and para-hydrogen, decay of excited states and photoselective photochemistry (using lamps - not yet lasers!). Theoretical chemistry played a central role in this endeavour. In 1943 the first book in Hebrew on quantum chemistry was written summarizing an extensive undergraduate and graduate seminar on this subject attended by mathematicians, physicists and chemists. In the early fifties theoretical chemistry became a part of the undergraduate curriculum at the Hebrew University, and in 1955 a book of the beautiful lectures of the late Aharon Katzir on quantum chemistry was published in the Hebrew language. This book constituted a tutorial text to many of our young scientists and also served as a language textbook for quantum chemists, who had immigrated to Israel and had joined our scientific community. In 1952 Professor Bernard Pullman and Professor Alberte Pullman, while visiting the Weizmann Institute of Science, delivered a series of lectures on theoretical chemistry, which constituted the first presentation of the methods of modern quantum chemistry to an audience of Israeli scientists. In 1962 the Hebrew University organized its first summer school on theoretical chemistry directed by the late Charles Coulson, which was held here on the Givat Ram Campus. This school can be considered as the symbolic onset of research activity in modern theoretical chemistry in Israel. During the last twenty-five years theoretical chemistry played a central role in the development of modern science in Israel and in fostering strong connections between our scientific community and the world scientific community. We hope to perpetuate this tradition.

The 6th ICQC is devoted to modern trends and developments in theoretical chemistry, emphasizing the interplay between theory and experiment in the broad area of quantum molecular sciences. Theoretical science should contribute towards the unification of concepts and to the establishment of a broad conceptual framework. In this context, it may be appropriate to dwell briefly on the conceptual framework of theoretical chemistry. This discipline in its broader sense, encompassing theoretical chemistry, chemical physics and biophysics, has undergone a metamorphosis during the last two decades, which I would like to characterize by three ingredients. Firstly, removal of boundaries: the central subject matter of traditional theoretical chemistry, chemical binding and structure, which constituted a major component of quantum chemistry until the sixties, has been blended with chemical dynamics and has emerged as a central ingredient of theoretical activity. Secondly, new areas: a multitude of novel research areas have joined theoretical chemistry. These include spectroscopy from small to huge molecules, molecular dynamics, the chemical reaction theory, solid state and surface phenomena, just to mention a few examples. Thirdly, new horizons: unification of concepts always constitutes a noble goal. Theoretical chemistry is becoming truly interdisciplinary, encompassing areas such as molecular astrophysics and biophysics. The 6th ICQC will reflect on these modern trends in theoretical chemistry.

I would like to express our sincere thanks to the distinguished scientific bodies, institutions and personalities that have made this Congress possible. The 6th ICQC is held under the auspices of the International Academy of Quantum Molecular Science. I would like to express my deep appreciation to the distinguished members of the Academy and its distinguished president for their sage advice and perceptive attitude in the planning of the programme of this Congress. The continuous advice we have received from the members of the Local Organizing Committee is also greatly appreciated. The Israeli universities have joined forces in support of this Congress. We are indebted to the Hebrew University, the Technion - Israel Institute of Technology, the Weizmann Institute of Science, Tel Aviv University and Bar-Ilan University for their support. This is appropriate for our university system which integrates higher education and basic research. I would like to convey special thanks and appreciation to the Hebrew University who graciously hosts our Congress. Thanks are due to the Israeli Ministry of Sciences and the National Research Council for their support. The 6th ICQC was most generously supported by the Fondation Edmond de Rothschild. We are greatly indebted to Baron Edmond de Rothschild for his support. I would like to express our gratitude and appreciation to the previous President of the International Academy of Quantum Molecular Science, Professor Bernard Pullman, for his remarkable help and assistance in this matter. We are greatly indebted to Baron Edmond de Rothschild for his magnificent involvement in our Congress, which will contribute significantly to science.

Our Congress is devoted to the conceptual framework of chemistry. Chemistry in Jerusalem is old. Two thousand nine hundred years ago in the construction of Solomon's Temple chemical techniques were used, utilizing methods of alloy production and of dye preparation. We shall not and will not be distracted from science, but let us remember that our Congress convenes in the city of Jerusalem; for Jerusalem reflects the continuous hisorical development, over a time span of more than three thousand years, of moral and cultural values. I do hope very much that during this Congress we shall all absorb novel and exciting scientific knowledge, and that simultaneously we shall also be inspired by the universal cultural message conveyed by the unique city of Jerusalem.

I wish us all a stimulating and inspiring scientific meeting.

DENSITY FUNCTIONAL THEORY

MEL LEVY
Chemistry Department and Quantum Theory Group
Tulane University
New Orleans, Louisiana 70118
U.S.A.

ABSTRACT. The formalism of density-functional theory for electronic structure is reviewed. Simplified proofs of the Hohenberg-Kohn existence theorems are given for ground states by employing the new constrained-search reformulation of density-functional theory. It is brought out that the exact correlation energy is a functional of the Hartree-Fock density. The status of excited-state theories are reviewed with focus upon what is and what is not allowed. The constrained-search approach and coordinate scaling are used to reveal the general form of F[n] which is the universal ground-state functional for the kinetic plus electron-electron repulsion energy for each trial electron density $n(\vec{r})$. The non-interacting kinetic and exchange components of F[n] exhibit homogeneous scaling but the correlation component exhibits complicated scaling. Explicit approximate functionals are displayed and studied. It is then pointed out that an approximate F[n] may yield in principle exact results for a range of external potentials. Finally, the Hartree-like Kohn-Sham equations are analyzed with emphasis on the highest-occupied orbital energy as the negative of the exact ionization energy.

1. Introduction

The origins of density-functional theory can be traced to Thomas-Fermi theory [1,2]. Modern density-functional theory stems from the Hohenberg-Kohn theorems [3] which imply a drastic simplification of the many-electron problem because the electron density, $n(\vec{r})$, is only 3-dimensional independent of the size of the system under consideration. It is the purpose of this lecture to discuss the fundamental formalism of density-functional theory. For simplicity of presentation, I shall restrict my discussion to this theory with the understanding that the results may easily be extended to the more general spin-density functional theory [4,5]. I shall make extensive use of the new and simple "constrained-search" reformulation [6,7] of the Hohenberg-Kohn theorems.

J. Jortner and B. Pullman (eds.), Perspectives in Quantum Chemistry, 1–17.

2. Hohenberg-Kohn Theorems for Ground States

2.1 CONSTRAINED-SEARCH PROOF OF THEOREM I

Before discussing the first Hohenberg-Kohn theorem, it should be pointed out that an infinite number of wavefunctions yield any $n(\vec{r})$. Consider, for example, a one-electron density. All the wavefunctions of the form $\Psi(\vec{r}) = n^{1/2}(\vec{r}) \exp[if(\vec{r})]$ generate the same $n(\vec{r})$; $\Psi^*(\vec{r})\Psi(\vec{r}) = n(\vec{r})$, for arbitrary real $f(\vec{r})$. The generalization by Harriman to many-electron densities is straightforward [8].

Consider N interacting electrons in a local spin-independent external potential v. The Hamiltonian is

$$H = T + V_{ee} + \sum_{i=1}^{N} v(\vec{r}) , \tag{1}$$

where T and V_{ee} are, respectively, the kinetic and electron-electron repulsion operators. According to Hohenberg-Kohn Theorem I, a given ground-state n_o determines its ground-state wavefunction Ψ_o, and consequently all the ground-state and excited-state properties of H, even though there are an infinite number of Ψ's which give n_o.

By means of the constrained-search orientation, the proof of Hohenberg-Kohn theorem I becomes transparent. One simply asks, "what is so special about Ψ_o?" Well, Ψ_o is that antisymmetric function which is constrained to yield n_o and simultaneously minimizes $\langle T+V_{ee}\rangle$. So, we see immediately how n_o determines Ψ_o. Moreover, since Ψ_o cannot be an eigenstate of more than one H, it follows that n_o determines H; change H by more than an additive constant in v and n_o must change. In short

$$n_o \longrightarrow \Psi_o \longrightarrow H . \tag{2}$$

Also, it should be clear that if the minimum in $\langle T+V_{ee}\rangle$ is achieved by more than one antisymmetric function, then each minimizing function must give the same expectation value with respect to any H, so it follows that each minimizing function is a ground-state of the same H. Thus, again $n_o \to H$, and degeneracies present no problems.

2.1.1. *Hartree-Fock Density Determines Everything*. The Hohenberg-Kohn Theorem I has been extended to the Hartree-Fock density [9-11], n_{HF}, in a rigorous manner [11] through the Fock Hamiltonian [11]. Consequently n_{HF} contains all the information about H, including the correlation energy [10,11]. That is,

$$n_{HF} \longrightarrow v \longrightarrow (E_o - E_{HF}) = E_c[n_{HF}]. \tag{3}$$

The existence of the universal correlation energy functional $E_c[n_{HF}]$ provides hope for a practical addition of $E_c[n_{HF}]$ to E_{HF} after one has performed a traditional Hartree-Fock calculation for E_{HF} and n_{HF}; one does not need the exact n_o! Namely,

$$E_o = E_{HF} + E_c\,[n_{HF}]. \tag{4}$$

2.2. CONSTRAINED-SEARCH PROOF OF THEOREM II

Hohenberg-Kohn Theorem II states that there exists a universal F[n] such that

$$E_o = \min_n \{\int v(\vec{r})n(\vec{r}) + F[n]\}. \quad (5)$$

F[n] is universal in that its value is independent of the particular v of interest. Again, the constrained-search formulation provides a direct proof. A proof by construction is employed [6,7]. Simply define F[n] to be [6,7]

$$F[n] = \min_{\Psi\to n} < \Psi|T+V_{ee}|\Psi > \quad (6)$$

or

$$F[n] = < \Psi_n^{min}|T+V_{ee}|\Psi_n^{min} >, \quad (7)$$

where Ψ_n^{min} minimizes $<T+V_{ee}>$ and is constrained to yield the trial n. Eq. (5) follows directly because

$$\int v(\vec{r})n(\vec{r})d\vec{r} + F[n]$$

$$= \min_{\Psi\to n} < \Psi|T+V_{ee} + \sum_{i=1}^{N} v(\vec{r}_i)|\Psi >$$

$$= \min_{\Psi\to n} < \Psi|H|\Psi > \quad (8)$$

implies

$$\min_n \{\int v(\vec{r})n(\vec{r})d\vec{r} + F[n]\}$$

$$= \min_n \min_{\Psi\to n} < \Psi|H|\Psi >$$

$$= \min_{\Psi} < \Psi|H|\Psi > = E_o. \quad (9)$$

Harriman [8] has shown that N-representability is no problem for any trial n and Lieb [12] has shown that Ψ_n^{min} always exists. In addition, Valone [13] has provided an ensemble-search generalization to Eqs. (6) and (7). The definition of F[n] in Eqs. (6) and (7) and its ensemble-search generalization follow in the spirit of Percus' [14] definition of a universal kinetic energy functional for independent fermion systems.

3. Excited-State Theories

Except for the lowest state of a given symmetry, the above ground-state theorems do not quite carry over to individual excited states. In particular, there exists no universal $F_M[n]$ for the variational determination [15,16] of individual state M, M>0. Instead, there exists an $F_M[n]$ such that

$$E_M \geq \min_n \{\int v(\vec{r})n(\vec{r})d\vec{r} + F_M[n]\}, \qquad (10)$$

with the inequality expected [16].

There does exist, however, a universal functional which gives E_M if one first finds n_o, then n_1, then n_2, ... then n_{M-1}. As an example, the functional for M=1 is $F[n_A,n_B]$. $F[n_A,n_B]$ first finds that antisymmetric wavefunction, $\Psi_{n_A}^{min}$, which yields n_A and minimizes $<T+V_{ee}>$. Then $F[n_A,n_B]$ finds Ψ_{n_A,n_B}^{min} which yields n_B, is orthogonal to $\Psi_{n_A}^{min}$, and minimizes $<T+V_{ee}>$. $F[n_A,n_B]$ is thereby defined by [15]

$$F[n_A,n_B] = <\Psi_{n_A,n_B}^{min}|T+V_{ee}|\Psi_{n_A,n_B}^{min}> \qquad (11)$$

and it follows that [15]

$$E_1 = \min_n \{\int v(\vec{r})n(\vec{r})d\vec{r} + F[n_o,n]\}. \qquad (12)$$

Theophilou has defined a functional $F_{0,1}[n]$ which generates $E_0 + E_1$ directly. Specifically [17],

$$E_o + E_1 = \min_n \{\int v(\vec{r})n(\vec{r})d\vec{r} + F_{0,1}[n]\}, \qquad (13)$$

with

$$F_{0,1}[n] = \min_{(\Psi',\Psi'')} [<\Psi'|T+V_{ee}|\Psi'> + <\Psi''|T+V_{ee}|\Psi''>], \qquad (14)$$

where $<\Psi'|\Psi''> = 0$ and $\int|\Psi'|^2 + \int|\Psi''|^2 \rightarrow n$. Recently, Gross, Oliveira, and Kohn [18] have extended Theophilou's excited-state development and they have actually produced encouraging numbers for simple systems [19].

4. Partitioning of the Ground-State F[n]

For the remainder of the lecture I shall focus upon the ground-state theory. What one desires is an accurate approximation of F[n] for computational purposes. To this end, it is convenient to partition F[n] according to [20]

$$F[n] = T_s[n] + U[n] + E_{xc}[n], \tag{15}$$

where

$$T_s[n] = < \Phi_n^{min} | T | \Phi_n^{min} >, \tag{16}$$

where U[n] is the classical electron-electron repulsion energy, where Φ_n^{min} is that single determinant which yields n and minimizes just $<T>$, and where $E_{xc} = E_x + E_c$ is the exchange-correlation functional. The exchange component, E_x, is normally defined as [21]

$$E_x[n] = < \Phi_n^{min} | V_{ee} | \Phi_n^{min} > - U[n] \tag{17}$$

which gives the correlation component as [21]

$$E_c[n] = < \Psi_n^{min} | T+V_{ee} | \Psi_n^{min} > - < \Phi_n^{min} | T+V_{ee} | \Phi_n^{min} >. \tag{18}$$

5. Coordinate Scaling of F[n]

The general explicit forms of $T_s[n]$, $E_x[n]$, and $E_c[n]$, may be ascertained by coordinate scaling. With this in mind, use shall be made of the fact that if $\Psi_n^{min,\alpha}$ yields n and minimizes $<T+\alpha V_{ee}>$ then, by Levy and Perdew [22], $\beta^{3N/2}\Psi_n^{min,\alpha}(\beta\vec{r}_1, \beta\vec{r}_2, \ldots, \beta\vec{r}_N)$ minimizes $<T+\alpha\beta V_{ee}>$ and yields the scaled density n_β, where $n_\beta(\vec{r}) = \beta^3 n(\beta\vec{r})$. (Note that most often $\Phi_n^{min} = \Psi_n^{min,0}$ and I shall assume this to be the case unless otherwise indicated. Also, I shall supress the spin coordinates to simplify notation).

5.1 COORDINATE SCALING DETERMINES THE FORMS OF T_s AND E_x

Since $\Phi_n^{min} = \Psi_n^{min,0}$, Φ_n^{min} minimizes just $<T>$, and it follows that $\lambda^{3N/2}\Phi_n^{min}(\lambda\vec{r}_1, \ldots, \lambda\vec{r}_N)$ is that single determinant which yields n_λ and minimizes just $<T>$. In other words,

$$\Phi_{n_\lambda}^{\min}(\vec{r}_1, \ldots, \vec{r}_N) = \lambda^{3N/2}\Phi_n^{\min}(\lambda\vec{r}_1, \ldots, \lambda\vec{r}_N). \tag{19}$$

This means that [22]

$$T_s[n_\lambda] = \lambda^2 T_s[n] \tag{20}$$

and

$$E_x[n_\lambda] = \lambda E_x[n], \tag{21}$$

where it has been taken into consideration that T is homogeneous of degree -2 and V_{ee} is homogeneous of degree -1.

According to Eq. (20) the local density component of T_s must have the form $\int n^{5/3}$ (Thomas-Fermi form), and according to Eq. (21) the local density component of E_x must have the form $\int n^{4/3}$ (Dirac form). The forms of nonlocal improvements upon $\int n^{5/3}$ and $\int n^{4/3}$ are also dictated, to a significant extent, by Eqs. (20) and (21). For instance [23],

$$\begin{aligned}\tilde{T}_s[n] &= a(N) \int n^{5/3}(\vec{r})d\vec{r} \\ &+ b(N) \int n^{-1}|\vec{\nabla}n|^2 d\vec{r} \\ &+ c(N) \int n(\vec{r})\vec{r}\cdot\vec{\nabla}[n^{-2/3}|\nabla n|]d\vec{r},\end{aligned} \tag{22}$$

where the second term is von-Weizsäcker-like and the third term is due to Levy and Oui-Yang in order to produce a correct property of the functional derivative. Each term in Eq. (22) scales as λ^2. (The tilde is used to signify that the functional is an approximate one.)

One of the powers of coordinate scaling is the fact that it is most often quite easy to observe immediately how a functional scales. Indeed, to transform any functional G[n] into $G[n_\lambda]$, one simply divides G[n] by λ^3, one replaces each n by $\lambda^3 n$, one replaces each $\vec{r}$ by $\lambda^{-1}\vec{r}$, and one replaces each $\vec{\nabla}$ by $\lambda\vec{\nabla}$. So, for example, $\nabla^2 n/n^{4/3}$ becomes $\lambda\nabla^2 n/n^{4/3}$ and $|\vec{\nabla}n|/n^{4/3}$ remains unchanged.

Attractive nonlocal exchange functionals now include the one by Perdew and Wang [24]:

$$\tilde{E}_x[n] = -\beta\int d\vec{r} n^{4/3}(\vec{r})Q(S), \tag{23}$$

$$Q(S) = (1+aS^2 + bS^4 + cS^6)^{1/15}, \tag{24}$$

where $S = \gamma|\vec{\nabla}n|/n^{4/3}$, and the one by Ghosh and Parr [25]:

$$\tilde{E}_x[n] = \omega\int d\vec{r} n^3(\vec{r})t(\vec{r},n)^{-1}, \quad (25)$$

where

$$t(\vec{r},n) = \frac{1}{8}\sum_i (\vec{\nabla}n_i\cdot\vec{\nabla}n_i)n_i^{-1} - \frac{1}{8}\nabla^2 n. \quad (26)$$

Both $\tilde{E}_x$ functionals, Eq. (23) and Eq. (25), satisfy condition (21). Note that the local density approximation (LDA) for E_x results from Eq. (23) if F(S) is set equal to unity and the constant β is adjusted. To gauge the evolution of the quality of $\tilde{E}_x$, various values for $\tilde{E}_x$ are displayed in Table I. Both nonlocal functionals give results which are significantly better than the LDA. But, it is well known that the LDA error in the exchange is, in part, offset by the LDA error in the correlation energy.

Table I. Values[a] for $\tilde{E}_x[n]$.

Atom	LDA[b]	Ghosh-Parr[c]	Perdew-Wang[d]	Exact[e]
He	0.884	0.913	1.033	1.026
Ar	27.86	29.24	30.29	30.18
Xe	170.6	181.7	178.6	179.1

[a]Atomic units are used.

[b]Local density approximation.

[c]Nonlocal Ref. 25.

[d]Nonlocal. Ref. 24.

[e]Hartree-Fock.

5.2 COORDINATE SCALING AND THE FORM OF E_c

Unlike $T_s[n]$ and $E_x[n]$, the correlation energy, $E_c[n]$, does not satisfy homogeneous scaling. Instead, E_c exhibits complicated scaling [22]. The reason [22] for this is the fact that

$$\Psi_{n_\lambda}^{min} \neq \lambda^{3N/2}\Psi_n^{min}(\lambda\vec{r}_1, \ldots, \lambda\vec{r}_N) \quad (27)$$

because the scaled wavefunction $\lambda^{3N/2}\Psi_n^{min}(\lambda\vec{r}_1, \ldots, \lambda\vec{r}_N)$ minimizes $\langle T+\lambda V_{ee}\rangle$ and <u>not</u> $\langle T+V_{ee}\rangle$. As a result, E_c satisfies the following scaling inequalities:

$$E_c[n_\lambda] > \lambda E_c[n]; \quad \lambda>1 \tag{28}$$

and

$$E_c[n_\lambda] < \lambda E_c[n]; \quad \lambda<1. \tag{29}$$

In addition, E_c scales differently at high λ than at low λ.

Quite recently [26], the asymptotic scaling requirements of E_c have been derived by observing that

$$E_c[n_\lambda] = < \Psi^{min}_{n_\lambda}|T+V_{ee}|\Psi^{min}_{n_\lambda} > - < \Phi^{min}_{n_\lambda}|T+V_{ee}|\Phi^{min}_{n_\lambda} > \tag{30}$$

implies

$$E_c[n_\lambda] = \lambda^2[< \Psi^{min,\alpha}_n|T|\Psi^{min,\alpha}_n > - < \Phi^{min}_n|T|\Phi^{min}_n >]$$

$$+ \lambda[< \Psi^{min,\alpha}_n|V_{ee}|\Psi^{min,\alpha}_n > - < \Phi^{min}_n|V_{ee}|\Phi^{min}_n >] \tag{31}$$

because

$$\Psi^{min}_{n_\lambda} = \lambda^{3N/2}\Psi^{min,\alpha}_n(\lambda\vec{r}_1, \ldots, \lambda\vec{r}_N), \tag{32}$$

where $\alpha = \lambda^{-1}$. Further, since $\Psi^{min,\alpha}_n$ minimizes $<T+\alpha V_{ee}>$ it follows that [22] $[< \Psi^{min,\alpha}_n|T|\Psi^{min,\alpha}_n > - < \Phi^{min}_n|T|\Phi^{min} >]$ is never negative and increases monotonically with increasing α, and $[< \Psi^{min,\alpha}_n|V_{ee}|\Psi^{min,\alpha}_n > - < \Phi^{min}_n|V_{ee}|\Phi^{min}_n >]$ is never positive and decreases monotonically to a constant as α increases.

Now, expand $\Psi^{min,\alpha}_n$ in the following perturbation series:

$$\Psi^{min,\alpha}_n(\vec{r}_1, \ldots, \vec{r}_N) = \Phi^{min}_n(\vec{r}_1, \ldots, \vec{r}_N) + $$

$$+ \sum_{K=1}^{\infty} \alpha^K g_K(\vec{r}_1,\ldots,\vec{r}n) \tag{33}$$

which applies for small α. Substitute Eq. (33) into Eq. (31) to yield

$$E_c[n_\lambda] = a[n] + b[n]\lambda^{-1} + c[n]\lambda^{-2} + \ldots \tag{34}$$

which gives

$$\lim_{\lambda\to\infty} E_c[n_\lambda] > -\infty. \tag{35}$$

In the other direction, as $\lambda\to 0$, $\alpha\to\infty$ so that $<\Psi_n^{min,\alpha}|V_{ee}|\Psi_n^{min,\alpha}> - <\Phi_n^{min}|V_{ee}|\Phi_n^{min}>$ approaches a negative constant. This means that

$$\lim_{\lambda\to 0} \lambda^{-1} E_c[n_\lambda] = -g[n], \tag{36}$$

where $g[n] > 0$. Further, since $<\Psi_n^{min,\alpha}|V_{ee}|\Psi_n^{min,\alpha}> \geq 0$ it follows that

$$\lim_{\lambda\to 0} \lambda^{-1} E_c[n_\lambda] \geq -<\Phi_n^{min}|V_{ee}|\Phi_n^{min}> \tag{37}$$

or

$$\lim_{\lambda\to 0} \lambda^{-1} E_c[n_\lambda] \geq -U[n] - E_x[n] > -U[n] \tag{38}$$

which means that

$$0 < g[n] \leq U[n] + E_x[n] < U[n]. \tag{39}$$

Expression (27) also leads to

$$T[n_\lambda] + V_{ee}[n_\lambda] < \lambda^2 T[n] + \lambda V_{ee}[n] \tag{40}$$

for $\lambda\neq 1$, where $T[n] = <\Psi_n^{min}|T|\Psi_n^{min}>$ and $V_{ee}[n] = <\Psi_n^{min}|V_{ee}|\Psi_n^{min}>$. It has been stated incorrectly [27] that the constrained-search density functionals do not obey the virial theorem because Eq. (40) is an inequality and not the equality which appears in wavefunctional theory. But, the constrained-search density functionals <u>do</u> obey the virial theorem because the inequality in Eq. (40) is always in the same direction!

As stated earlier, the homogeneous scaling relations (20) and (21) dictate very simple forms for T_s and E_x within the local density approximation. Namely, the LDA approximation for T_s is proportional to $\int n^{5/3}$ and the LDA approximation for E_x is proportional to $\int n^{4/3}$. The simplest form which satisfies inequalities (28) and (29), which is bounded as $\lambda\to\infty$, and which goes as λ when $\lambda\to 0$, is the well-known Wigner correlation functional [28]

$$\tilde{E}_c[n] = -a\int n(\vec{r})\left[b+cn(\vec{r})^{-1/3}\right]^{-1} d\vec{r}, \tag{41}$$

where the constants a, b, and c are greater than zero. This functional is more complicated than its simplest possible $\tilde{T}_s$ and $\tilde{E}_x$ counterpart because the scaling conditions are more complicated. The

Wigner functional is a special case of

$$\tilde{E}_c[n_\lambda] = -\int A([n];\vec{r})[\lambda^{-1}B([n];\vec{r}) + D([n];\vec{r})]^{-1}n(\vec{r})d\vec{r}, \quad (42)$$

where the A, B, and D may be local or nonlocal. Eq. (42) is going to be investigated in the future. Note that the Wigner $\tilde{E}_c$ goes unreasonably, as $\lambda\to\infty$, to the same constant, $-ab^{-1}N$, regardless of n. This is not true of Eq. (42).

The local density $\tilde{E}_c$ is defined by

$$\tilde{E}_c[n] = \int n(\vec{r})\epsilon_c^{LDA}[n(\vec{r})]d\vec{r}, \quad (43)$$

where $\epsilon_c^{LDA}[n(\vec{r})]$, at point $\vec{r}$, is the correlation energy per electron of a uniform electron gas of density $n(\vec{r})$. The Gunnarsson-Lundqvist parameterization [29] for ϵ_c^{LDA} is

$$\epsilon_c^{LDA}[n(\vec{r})] = -aG(x), \quad (44)$$

where

$$G(x) = \frac{1}{2}\left[(1+x^3)\ln(1+x^{-1}) - x^2 + \frac{x}{2} - \frac{1}{3}\right], \quad (45)$$

and where $x = bn^{-1/3}$. Eqs. (43-45) give more reasonable correlation energies than Eq. (41).

Further improvements in $\tilde{E}_c$ have been obtained by means of nonlocal functionals. Recent attractive examples include the following one by Lee, Yang, and Parr [30]:

$$\tilde{E}_c[n] = -a\int(1+dn^{-1/3})^{-1}\left[n+bn^{-2/3}w(n)\exp(-cn^{-1/3})\right] \quad (46)$$

where

$$w(n) = c_F n^{5/3} - 2t_w + (\tfrac{1}{9}t_w + \tfrac{1}{18}\nabla^2 n), \quad (47)$$

and where

$$t_w = \tfrac{1}{2}|\nabla n|^2 n^{-1} - \tfrac{1}{8}\nabla^2 n, \quad (48)$$

and the following one of Perdew [31]:

$$\tilde{E}_c[n] = \int d\vec{r}n(\vec{r})\epsilon_c^{LDA}[n(\vec{r})] + \int d\vec{r}\, e^{-\Phi}G(n)|\nabla n|^2 n^{-2/3}, \quad (49)$$

where

$$\Phi = a|\nabla n|^{-7/6}, \quad (50)$$

where

$$G(n) = b + (\alpha+\beta r_s+\gamma r_s^2)(1+tr_s+ur_s^2+wr_s^3)^{-1}, \tag{51}$$

and where $r_s = cn^{-1/3}$. Perdew's functional is an improvement upon the Langreth-Mehl [32] non-local $\tilde{E}_c$. Errors from the LDA and from the non-local $\tilde{E}_c$'s, Eqs. (46) and (49), are given in Table II. Note the improvement made by the nonlocal functional as compared with the LDA. Of course, it should be noted that the LDA error in correlation is cancelled, in part, by the LDA error in exchange.

Future further improvement in approximations to E_c should incorporate the fact that neither the LDA, Eq. (43), nor the above nonlocal functionals, Eqs. (46) and (49), satisfy the bounded $\lambda\to\infty$ condition, Eq. (35). These functionals are unbounded as $\lambda\to\infty$. The Lee-Yang-Parr functional goes as $-\lambda^2$ and both the Perdew functional and the LDA go as $-\ln\lambda$. Also, the Langreth-Mehl functional actually goes as $+\lambda$ and thus becomes $+\infty$.

Table II. Errors[a] in $\tilde{E}_c[n]$

Species	LDA[b]	Perdew[c]	Lee-Yang-Parr[d]
He	0.07	0.003	0.002
N	0.24	0.017	0.004
H_2O	0.29	-0.007	0.034
Xe	0.3	0.035	0.001

[a] Atomic units used.

[b] Local density approximation.

[c] Nonlocal. Ref. 31.

[d] Nonlocal. Ref. 30.

6. Kohn-Sham Hartree-Like Equations

Most practical density-functional calculations on solids and large molecules circumvent approximating T_s by evaluating the latter explicitly. In particular, according to Kohn-Sham theory [20], the minimization in Eq. (5) results in

$$[-\tfrac{1}{2}\nabla^2 + v + u + v_{xc}]\phi_i^{KS} = \epsilon_i\phi_i^{KS}, \quad i = 1, 2, \ldots, N \qquad (52)$$

where $v_{xc}(\vec{r}) = \delta E_{xc}/\delta n$ and $u(\vec{r}) = \delta U/\delta n$, so that $T_s[n_o] = \sum_{i=1} < \phi_i^{KS}| - \frac{1}{2}\nabla^2|\phi_i^{KS} >$. The Kohn-Sham Hartree-like equations, Eqs. (52), are quite accurate because the effective potential $v+u+v_{xc}$ is local and therefore Hartree-like. As a result, self-consistency is achieved more easily than in Hartree-Fock theory and Eq. (52) generates the exact ground-state density and energy when the exact E_{xc} is employed, even though only a single determinant is implicitly used.

One surprising feature of Eq. (52) is the fact that [33-36]

$$\epsilon_N = - I, \qquad (53)$$

with $\lim_{|\vec{r}|\to\infty} [v(\vec{r}) + u(\vec{r}) + v_{xc}(\vec{r})] = 0$, where ϵ_N is the highest-occupied orbital energy and I is the <u>exact</u> ionization energy. We have also recently learned, however, that $(\epsilon_N - \epsilon_{N+1})$ underestimates [37-38] the band gap in insulators and semiconductors.

Eq. (53) arises from the fact that

$$- \tfrac{1}{2}\nabla^2\phi_i(\vec{r}) = \epsilon_i\phi_i(\vec{r}); \ |\vec{r}|\to\infty, \qquad (54)$$

implies

$$|\phi_i(\vec{r})|^2 \sim \exp[- 2(2|\epsilon_i|)^{1/2}|\vec{r}|]; \ |\vec{r}|\to\infty. \qquad (55)$$

Also, since

$$n(\vec{r}) = \sum_{i=1}^{N} |\phi_i(\vec{r}_i)|^2 \qquad (56)$$

and $|\epsilon_N| \le |\epsilon_i|$, it follows that [39]

$$n(\vec{r}) = |\phi_N(\vec{r})|^2; \ |\vec{r}|\to\infty. \qquad (57)$$

Further, we already know that [35,40-45]

$$n(\vec{r}) \sim \exp[- 2(2I)^{1/2}|\vec{r}|]; \ |\vec{r}|\to\infty. \qquad (58)$$

Finally, comparison of Eqs. (58), (57), and (55) with i = N, results in Eq. (53).

Eq. (53) is surprising because the exact ionization energy is generated from a non-interacting system of equations and because actual calculations have often yielded an approximate $|\epsilon_N|$ which is far from I even though the approximate n is often reasonably accurate. This latter point is explained by the fact that ϵ_N is very sensitive to the "tail" of $n(\vec{r})$, and an otherwise good density may have a poor tail. Consider, as a simple example, the H-atom where of course

$$\left[-\frac{1}{2}\nabla^2 - \frac{1}{r}\right]e^{-r} = (-0.50)\, e^{-r}. \tag{59}$$

Now we can change the eigenfunction (and density) by an <u>imperceptible amount</u> and change the eigenvalue <u>drastically.</u> For instance, consider

$$\left[-\frac{1}{2}\nabla^2 - w(\vec{r})\right](e^{-r} + 10^{-100}e^{-0.4r})$$

$$= -0.08\,(e^{-r} + 10^{-100}e^{-0.4r}), \tag{60}$$

where $w(\vec{r})\to 0$ as $|\vec{r}|\to\infty$. Except at very large $|\vec{r}|$, at the tails of the n's, the potential $w(\vec{r})$ differs from r^{-1} by only a constant equal to the large difference in the two eigenvalues (0.50–0.08).

7. On the Present Status of Computational Density-Functional Theory

In this lecture I have focused upon the formalism of density-functional theory rather than upon its computational status. General reviews of density functional theory are forthcoming in a book by Parr and Yang and in a "Reviews of Modern Physics" article by Jones and Gunnarsson. A recent article by Kutzler and Painter [46], which studies simple systems, illustrates aspects of the present status of calculations. They employed the Kohn-Sham procedure. Diatomic binding energies from the Kutzler- Painter article are reported in Table III where the well-known overbinding of the local spin-density approximation is evident even though the local spin-density approximation tends to give good geometries. Observe the binding energy improvement made with the use of nonlocal functionals.

Table III. Dissociation Energies[a] (electron volts).

Molecule	LDA[b]	Experimental	Nonlocal[c]
B_2	3.8	2.9	3.2
C_2	7.2	6.2	6.2
O_2	7.5	5.2	5.9
F_2	3.4	1.6	2.2

[a] As reported in Ref. 46.

[b] Local density approximation.

[c] Ref. 24 for $\tilde{E}_x$ and Ref. 31 for $\tilde{E}_c$.

8. Closing Remarks

The exact F[n] yields the exact ground-state energy and density for all external potentials and all N. There is consequently enormous interest in F[n] because, as stated in the Introduction, n always contains only three spatial coordinates, independent of N, in contrast to the wavefunction which contains 3N spatial coordinates. However, although F[n] can be defined very explicitly in a conceptually useful and generalized form by the "constrained-search" formulation, it is obvious that it is impossible to ever put down the exact F[n] in a computationally useful form. It is natural then to ask if, in principle, a functional has to be the exact F[n] to yield, upon energy minimization, the exact energies and densities for a given set of external potentials and a given set of electron numbers of interest. The answer, fortunately, is no. As a very simple illustrative example [47], consider

$$F'[n] = F[n] + \min_{(p,w)} \int [n(\vec{r}) - n_w^p(\vec{r})]^2 d\vec{r}, \tag{61}$$

where n_w^p is the ground-state density for p electrons and external potential w. It is understood that each p is restricted to the set P and each w is restricted to the set W. With F'[n], the exact ground-state energy and density is obtained upon energy minimization, whenever both $v \in W$ and $N \in P$, even though the wrong energy and density is obtained if either v is not within W or N is not within P. On the other side of the coin, the existence of good numerical results for a class of external potentials and electron number should never be the sole criterion for the general utility of a given approximate functional because the functional might be an F' that just happens to give accurate results for this given class of potentials and electron number and inaccurate results outside this class and number. In short, I caution against excessive use of numerical comparisons in the evaluation of the progress of density-functional theory. Exclusive use of numerical comparisons tends to lead to a cycle of optimism followed by unexpected disappointment. The proper evolution of the field depends upon proper analysis with wavefunctions and many-body tools as well as upon numerical comparisons. Happily, the field appears to be progressing on both fronts.

9. References

1. L. H. Thomas, Proc. Camb. Phil. Soc. **23**, 542 (1927).
2. E. Fermi, Rend. Acad. Naz. Lincei **6**, 602 (1927).
3. P. Hohenberg and W. Kohn, Phys. Rev. **136**, B864 (1964).
4. U. von Barth and L. Hedin, J. Phys. C **5**, 1629 (1972).
5. A. K. Rajagopal and J. Callaway, Phys. Rev. B **7**, 1912 (1973).
6. M. Levy, Proc. Natl. Acad. Sci. (USA) **76**, 6062 (1979).
7. M. Levy, Bull. Amer. Phys. Soc. **24**, 626 (1979).
8. J. E. Harriman, Phys. Rev. A **24**, 680 (1981).
9. P. W. Payne, J. Chem. Phys. **71**, 490 (1979).
10. R. A. Harris and L. R. Pratt, J. Chem. Phys. **83**, 4024 (1985).
11. M. Levy, pgs. 479-498 in Density Matrices and Density Functionals, R. Erdahl and Density Functionals, R. Erdahl and V. H. Smith, Jr. (eds.), D. Reidel publishing company (1987).
12. E. H. Lieb, Int. J. Quantum Chem. **24**, 243 (1983).
13. S. M. Valone, J. Chem. Phys. **73**, 4653 (1980).
14. J. K. Percus, Int. J. Quantum Chem. **13**, 89 (1978).
15. M. Levy and J. P. Perdew, pgs. 11-30 in Density Functional Methods in Physics, R. M. Driezler and J. da Providencia (eds.), Plenum (1985).

16. E. H. Lieb, pgs. 31-80 in Density Functional Methods in Physics, R. M. Driezler and J. da Providencia (eds.), Plenum (1985).

17. A. K. Theophilou, J. Phys. C **12**, 5419 (1979).

18. E. K. U. Gross, L. N. Oliveira, and W. Kohn, Phys. Rev. A **37**, 2805, 2809 (1988).

19. L. N. Oliveira, E. K. U. Gross, and W. Kohn, Phys. Rev. A **37**, 2821 (1988).

20. W. Kohn and L. J. Sham, Phys. Rev. **140**, A1133 (1965).

21. V. Sahni, J. Gruenbaum, and J. P. Perdew, Phys. Rev. B **26**, 4371 (1982).

22. M. Levy and J. P. Perdew, Phys. Rev. A **32**, 2010 (1985).

23. M. Levy and H. Ou-Yang, Phys. Rev. A **38**, 625 (1988).

24. J. P. Perdew and Y. Wang, Phys. Rev. B **33**, 8800 (1986).

25. S. K. Ghosh and R. G. Parr, Phys. Rev. A **34**, 785 (1986).

26. M. Levy, unpublished.

27. See the addendum by E. Kryachko, I. Z. Petkov, and M. V. Stoitsov, Int. J. Quantum Chem. **34**, 307 (1988).

28. E. P. Wigner, Phys. Rev. **46**, 1002 (1934); pg. 94 in D. Pines, Elementary Excitations in Solids, W. A. Benjamin (1963).

29. O. Gunnarsson and B. I. Lundqvist, Phys. Rev. B **13**, 4274 (1976).

30. C. Lee, W. Yang, and R. G. Parr, Phys. Rev. B **37**, 785 (1988).

31. J. P. Perdew, Phys. Rev. B **33**, 8822 (1986).

32. D. C. Langreth and M. J. Mehl, Phys. Rev. B **28**, 1809 (1983).

33. J. P. Perdew, R. G. Parr, M. Levy, and J. L. Balduz, Jr., Phys. Rev. Lett. **49**, 1691 (1982).

34. C.-O. Almbladh and U. von Barth, Phys. Rev. B **31**, 3231 (1985).

35. M. Levy, J. P. Perdew, and V. Sahni, Phys. Rev. A **30**, 2745 (1984).

36. P. M. Laufer and J. B. Kreiger, Phys. Rev. A **33**, 1480 (1986).

37. J. P. Perdew and M. Levy, Phys. Rev. Lett. **51**, 1884 (1983).

38. L. J. Sham and M. Schlüter, Phys. Rev. Lett. **51**, 1888 (1983).

39. J. P. Perdew and A. Zunger, Phys. Rev. B **23**, 5048 (1981).

40. M. M. Morrell, R. G. Parr, and M. Levy, J. Chem. Phys. **62**, 549 (1975).

41. M. Levy and R. G. Parr, J. Chem. Phys. **64**, 2702 (1976).

42. J. Katriel and E. R. Davidson, Proc. Natl. Acad. Sci. (U.S.A.) **77**, 4403 (1980).

43. H. J. Silverstone, Phys. Rev. A **23**, 1030 (1981).

44. G. Hunter, Int. J. Quantum Chem. Symp. **9**, 311 (1975).

45. M. Levy, University of North Carolina, Chapel Hill, Technical Report, unpublished (1975).

46. F. W. Kutzler and G. S. Painter, Phys. Rev. Lett. **59**, 1298 (1987).

47. J. E. Osburn and M. Levy, Phys. Rev. A **35**, 3233 (1987).

INTERMOLECULAR FORCES FROM THE VIEWPOINT OF QUANTUM ELECTRODYNAMICS

D.P. Craig
Australian National University
Department of Chemistry, The Faculties
GPO Box 4, Canberra, ACT, 2601, AUSTRALIA

Abstract

The techniques of molecular quantum electrodynamics unify the theory of radiation-molecule and molecule-molecule coupling, the latter being treated as an interaction conveyed or mediated by the radiation field. The calculation of intermolecular effects is readily taken to all separation distances, electromagnetic retardation being included in a natural way. The methods themselves are established by wide application, and are correct in all cases tested. The methods are outlined and illustrated in three cases: the primitive resonance between identical molecules, one ground state and one excited; the effect on dispersive binding of an external radiation field; and the change in the chiroptical properties of a coupled pair of molecules, one chiral and one achiral.

1. Introduction

It is conventional to accept that correct theories of molecular structure and molecular properties came only after quantum mechanics was discovered in 1925, and then applied to valence bonds by Heitler and London in 1927.

However forces between molecules at distances outside the overlap distance were in part understood in classical theory. The obvious example is the coupling between molecules A and B with permanent electric dipole moments $\underset{\sim}{\mu}^A$ and $\underset{\sim}{\mu}^B$. The energy shift at separation $\underset{\sim}{R}$ has long been known to be

$$\Delta E = \frac{1}{4\pi\varepsilon_o R^3}\left\{ \underset{\sim}{\mu}^A . \underset{\sim}{\mu}^B - 3\left(\underset{\sim}{\mu}^A . \hat{\underset{\sim}{R}}\right)\left(\underset{\sim}{\mu}^B . \hat{\underset{\sim}{R}}\right) \right\}$$

$$= \frac{1}{4\pi\varepsilon_o R^3}\left\{ \mu_i^A \mu_j^B - 3\left(\mu_i^A \hat{R}_i\right)\left(\mu_j^B \hat{R}_j\right)\right\}$$

J. Jortner and B. Pullman (eds.), Perspectives in Quantum Chemistry, 19–39.

$$= \frac{\mu_i^A \mu_j^B}{4\pi\varepsilon_o R^3} \beta_{ij} \tag{1.1}$$

$$\text{and } \beta_{ij} = \delta_{ij} - 3\hat{R}_i \hat{R}_j$$

where $\hat{\underset{\sim}{R}}$ is the unit vector along $\underset{\sim}{R}$, δ_{ij} is the Kronecker delta, and repeated indices are to be summed over. ε_o is the vacuum permittivity.

It is less well-known that the dispersion interaction between two molecules was already known in classical theory. The line of the argument was as follows. The essential classical idea is that in any molecular charge distribution there are charge fluctuations, described as 'spontaneous', which can be expanded into combinations of dipole, quadrupole and higher order charge displacements. The leading term is the dipole fluctuation.

If μ_i^A is such an instantaneous electric dipole moment component in A it produces an electric field at B, located at $\underset{\sim}{R}_B$, with $\underset{\sim}{R}_B - \underset{\sim}{R}_A = \underset{\sim}{R}$, according to (1.2)

$$E_j = -\frac{\mu_i^A}{4\pi\varepsilon_o R^3} \left(\delta_{ij} - 3\hat{R}_i \hat{R}_j\right) \tag{1.2}$$

Moments are induced in B by these fields, proportional to the polarizability $\underset{\sim}{\alpha}^B$,

$$\mu_m^B = \alpha_{lm}^B E_l = -\frac{\alpha_{lm}^B \mu_i^A}{4\pi\varepsilon_o R^3} \beta_{li} \tag{1.3}$$

The coupling to $\underset{\sim}{\mu}^A$ of the induced moments gives an energy shift

$$\Delta E = -\frac{\mu_i^A \mu_j^A}{16\pi^2 \varepsilon_o^2 R^6} \alpha_{lm}^B \beta_{mj} \beta_{il} \tag{1.4}$$

Noting that the magnitude of $\mu_i^A \; \mu_j^A$ is proportional in the classical picture to α_{ij}^A we have

$$\Delta E \propto -\frac{\alpha_{ij}^A \alpha_{lm}^B}{16\pi^2 \varepsilon_o^2 R^6} \beta_{il} \beta_{mj} \tag{1.5}$$

After quantum mechanics the 'spontaneous fluctuations' became virtual transitions. A virtual transition is merely an artefact of perturbation theory and nothing very profound has happened. The other change is profound, namely that we are able to give a definition of polarizability in terms of molecular energy states and transition strengths between them, so that, with the help of quantum theory we can calculate polarizabilities and get calculated energies that can be compared with measurements. That was not possible with the classical expression.

The expression for the dispersion interaction energy between freely rotating (identical) molecules could then be written

$$\Delta E = -\frac{1}{24\pi^2\varepsilon_o^2R^6}\sum_{r,s}\frac{|\underset{\sim}{\mu}^{ro}|^2\ |\underset{\sim}{\mu}^{so}|^2}{E_{ro}+E_{so}} \qquad (1.6)$$

$\underset{\sim}{\mu}^{ro}$ is the transition dipole moment for the transition $r \leftarrow o$, and E_{ro} is the transition energy.

This is the correct expression for separations R small compared with the transition wavelengths $\hbar c/E_{ro}$. It is not correct at large separations, as was first shown by Casimir and Polder [1]. This marks the beginning of the development of theories taking account of retardation, and ultimately using the apparatus of quantum electrodynamics.

The starting point for a discussion of how quantum electrodynamics enters the problem is to note that the coupling between molecules is conveyed by the electric field due to virtual transitions. Such a field is propagated at a finite speed, i.e. the speed of light. But the interaction of this oscillating field with the target molecule is treated in (1.5) as if it were a static interaction. That is to say the propagation time from one molecule to the other is treated as so short that in classical language the phase information is irrelevant. The induced dipole moment in the target molecule is exactly in phase with the initial charge fluctuation, and the field from the target molecule is propagated back from the initiating molecule still in perfect phase matching. We are not surprised that this picture applies well at distances very small compared with an optical wavelength. No modification is needed when the propagation time of R/c is short. When it is larger the phase of the returning field does not match the driving field.

Casimir and Polder showed that if the finite speed of propagation of light is allowed for at distances comparable to the wavelength of visible light but still much less than a wavelength, say a few hundred Angstrom, the characteristic inverse sixth power dependence of energy on distance in the static model (1.5) is supplemented by an inverse seventh power dependence. As the distance becomes still greater the inverse sixth power is altogether lost and the inverse seventh power is the leading term, giving in this limit an interaction (1.6).

$$\Delta E = -\frac{23\hbar c}{64\pi^3\varepsilon_o^2}\ \frac{\alpha(A)\alpha(B)}{R^7} \qquad (1.7)$$

$\alpha(A)$ and $\alpha(B)$ are the molecular isotropic polarizabilities.

This new result was later verified experimentally by Tabor and Winterton [2] in measurements of the attractive forces between parallel mica plates.

2. Molecular quantum electrodynamics

In this method the interaction between molecules is conveyed or mediated by the electromagnetic field, which is taken to be a constituent of the dynamical system. The complete Hamiltonian for a system of interacting atoms or molecules, outside the region of overlapping electron clouds, is

$$H = H_{mol} + H_{rad} + H_{int} \tag{2.1}$$

H_{mol} is the Hamiltonian for the molecules, H_{rad} the Hamiltonian for the radiation, and H_{int} the coupling between particles and field. The system is conservative and overall energy conservation holds; energy gained by the molecules, for example in an absorption process, is lost by the radiation field.

H_{mol} is the usual free molecule hamiltonian. In these applications its eigenvalues and eigenfunctions are assumed known. The radiation field hamiltonian H_{rad} needs to be discussed. The microscopic field vectors, which take account of the 'graininess' of the distribution of point charges are $\underset{\sim}{e}$, the electric field vector, and $\underset{\sim}{b}$ the magnetic field vector. $\underset{\sim}{E}$ and $\underset{\sim}{B}$ are conventionally reserved for the macroscopic fields of continuous distributions of charge and current. H_{rad} is to refer to the electromagnetic field in charge-free space, for which Maxwell's equations are

$$\underset{\sim}{\nabla} \cdot \underset{\sim}{e} = 0 \tag{2.2}$$

$$\underset{\sim}{\nabla} \cdot \underset{\sim}{b} = 0 \tag{2.3}$$

$$\underset{\sim}{\nabla} \times \underset{\sim}{e} = -\frac{\partial \underset{\sim}{b}}{\partial t} \tag{2.4}$$

$$\underset{\sim}{\nabla} \times \underset{\sim}{b} = \frac{1}{c^2}\frac{\partial \underset{\sim}{e}}{\partial t} \tag{2.5}$$

Quantization of the Maxwell fields is most easily brought within the usual scheme if we introduce the vector and scalar potentials $\underset{\sim}{a}$ and ϕ, according to

$$\underset{\sim}{b} = \underset{\sim}{\nabla} \times \underset{\sim}{a} \tag{2.6}$$

$$\underset{\sim}{e} = -\frac{\partial \underset{\sim}{a}}{\partial t} - \nabla\phi \tag{2.7}$$

Use of (2.4) and (2.5) leads to the wave equation in $\underset{\sim}{a}$,

$$\nabla^2 \underset{\sim}{a} - \frac{1}{c^2} \frac{\partial^2 \underset{\sim}{a}}{\partial t^2} = 0 \tag{2.8}$$

with plane wave solutions

$$\underset{\sim}{a} = \underset{\sim}{a}_o \, e^{i(\underset{\sim}{k}.\underset{\sim}{r} - \omega t)} \tag{2.9}$$

$\underset{\sim}{a}_o$ is a constant amplitude factor, $\underset{\sim}{k}$ is the wave vector, $\omega = c|\underset{\sim}{k}|$ the frequency and $\underset{\sim}{r}$ the position vector. $|\underset{\sim}{k}| = k$ is the wave number, equal to the number of wavelengths per unit distance along $\underset{\sim}{k}$.

We see from (2.6) and (2.7) that the potentials $\underset{\sim}{a}$ and ϕ are not uniquely defined as integrals of the $\underset{\sim}{b}$ and $\underset{\sim}{e}$ fields. They are determined only up to an additive gauge (scalar) function χ. The pair of transformations

$$\left.\begin{aligned} \underset{\sim}{a} &\rightarrow \underset{\sim}{a} + \underset{\sim}{\nabla}\chi \\ \phi &\rightarrow \phi - \frac{\partial \chi}{\partial t} \end{aligned}\right\} \tag{2.10}$$

have no effect on the $\underset{\sim}{e}$ and $\underset{\sim}{b}$ fields, nor on Maxwell's equations. We are thus free to choose the gauge to suit problems to be studied. For work with chemical systems, i.e. slow particles in bound states, Coulomb gauge is best, defined by

$$\underset{\sim}{\nabla}.\underset{\sim}{a} = 0 \tag{2.11}$$

The vector potential is purely transverse, and the total Hamiltonian (2.1) contains the Coulomb potential between charged particles in the same way as in the ordinary quantum mechanics of atoms and molecules.

In the transition to quantum theory, normalisation can be accomplished on states in which the continuum of solutions (2.9) is reduced to a countable infinity by the technique of box normalisation. Acceptable values of $\underset{\sim}{k}$ are limited to those for which solutions have the same values on opposite faces of a cubic box of side L. The allowed components are, for integers n_i,

$$k_i = \frac{2\pi n_i}{L} \; ; \quad i = x,y,z \tag{2.12}$$

The allowed $\underset{\sim}{k}$ values define pairs of modes of the electromagnetic field; one for each of the orthogonal polarization directions in the plane normal to $\underset{\sim}{k}$.

The mode expansion of $\underset{\sim}{a}(\underset{\sim}{r},t)$ is the Fourier series (2.13),

$$\underset{\sim}{a}(\underset{\sim}{r},t) = \sum_{\underset{\sim}{k}} \left\{ \underset{\sim}{a}_{\underset{\sim}{k}}(t)e^{i\underset{\sim}{k}\cdot\underset{\sim}{r}} + \bar{\underset{\sim}{a}}_{\underset{\sim}{k}}(t)e^{-i\underset{\sim}{k}\cdot\underset{\sim}{r}} \right\} \tag{2.13}$$

where $\bar{\underset{\sim}{a}}_k$ is the complex conjugate of $\underset{\sim}{a}_k$.

Each Fourier component is of the form, in space and time dependence,

$$\underset{\sim}{a}_{\underset{\sim}{k}}(\underset{\sim}{r},t) \sim e^{i(\underset{\sim}{k}\cdot\underset{\sim}{r} - \omega t)} \tag{2.14}$$

describing a wave propagating along $\underset{\sim}{k}$ with velocity c. The $\underset{\sim}{e}$ and $\underset{\sim}{b}$ fields are parallel to, and perpendicular to, the $\underset{\sim}{a}$ field for each mode $\underset{\sim}{k}$.

The steps toward quantization begin with a classical Hamiltonian, expressed in terms of the vector potential $\underset{\sim}{a}$, and a variable canonically conjugate to it, namely the momentum density

$$\underset{\sim}{\Pi} = \varepsilon_o \frac{\partial \underset{\sim}{a}}{\partial t} = - \varepsilon_o \underset{\sim}{e} \tag{2.15}$$

The total energy in terms of $\underset{\sim}{e}$ and $\underset{\sim}{b}$ is

$$H = \frac{\varepsilon_o}{2} \int \left(\underset{\sim}{e}^2 + c^2\underset{\sim}{b}^2\right) d^3\underset{\sim}{r} \tag{2.16}$$

After a development that is given fully elsewhere [3], using mode expansions for $\underset{\sim}{a}$ as in (2.13) and for the conjugate momentum density, H can be written as a sum of harmonic oscillator hamiltonians (2.17)

$$H = \sum_{\underset{\sim}{k},\lambda} H_{\underset{\sim}{k},\lambda} = \frac{1}{2} \sum_{\underset{\sim}{k},\lambda} \left\{ p_{\underset{\sim}{k}}^{(\lambda)2} + \omega^2 q_{\underset{\sim}{k}}^{(\lambda)2} \right\} \tag{2.17}$$

where the (real) momenta $p_k^{(\lambda)}$ and displacements $q_k^{(\lambda)}$ are related to the amplitudes of the vector potential,

$$\left.\begin{aligned} q_{\underset{\sim}{k}}^{(\lambda)} &= \left(V\varepsilon_o\right)^{1/2} \left(a_{\underset{\sim}{k}}^{(\lambda)} + \bar{a}_{\underset{\sim}{k}}^{(\lambda)}\right) \\ p_{\underset{\sim}{k}}^{(\lambda)} &= -ick\left(V\varepsilon_o\right)^{1/2} \left(a_{\underset{\sim}{k}}^{(\lambda)} - \bar{a}_{\underset{\sim}{k}}^{(\lambda)}\right) \end{aligned}\right\} \tag{2.18}$$

where the superscript λ gives the polarization direction and $V = L^3$ is the quantization volume.

The Hamiltonian (2.17) is now quantized in the familiar way, with allowed energies $(n + 1/2)\hbar ck$ in each mode k, with n = 0,1,2,...

In second quantization we can write the Hamiltonian for the radiation as

$$H_{rad} = \sum_{\underset{\sim}{k},\lambda} \left\{ a^{\dagger(\lambda)}(\underset{\sim}{k}) a^{(\lambda)}(\underset{\sim}{k}) + \frac{1}{2} \right\} \hbar ck \tag{2.19}$$

where $a^{\dagger(\lambda)}(\underset{\sim}{k})$ and $a^{(\lambda)}(\underset{\sim}{k})$ are creation and annihilation operators for the mode $(\underset{\sim}{k},\lambda)$ with commutation properties

$$\left.\begin{aligned} \left[a^{(\lambda)}(\underset{\sim}{k}),\ a^{(\lambda')}(\underset{\sim}{k}')\right] &= \left[a^{\dagger(\lambda)}(\underset{\sim}{k}),\ a^{\dagger(\lambda')}(\underset{\sim}{k}')\right] = 0 \\ \left[a^{(\lambda)}(\underset{\sim}{k}),\ a^{\dagger(\lambda')}(\underset{\sim}{k}')\right] &= \delta_{\underset{\sim}{k}\underset{\sim}{k}'}\ \delta_{\lambda\lambda'} \end{aligned}\right\} \tag{2.20}$$

The commutation rules apply to Bose particles, the particles being the photons associated with the electromagnetic field. The states of the electromagnetic field are specified by kets, in which the numbers of photons (occupation numbers) are set out for the occupied modes, as in (2.21)

$$|n_1(\underset{\sim}{k}_1,\lambda_1),\ n_2(\underset{\sim}{k}_2,\lambda_2),\ n_3(\underset{\sim}{k}_3,\lambda_3),\ \ldots\rangle \tag{2.21}$$

We see that when all occupation numbers are zero, for the vacuum field, there remains a zero-point energy in each mode (see (2.19)). In this state the electric and magnetic fields do not disappear, but fluctuate about mean values. It can be shown that the mean value (2.22) of the electric field is zero, while the expectation of the square of the field (2.23) is nonzero,

$$\langle 0|\underset{\sim}{e}|0\rangle = 0 \tag{2.22}$$

$$\langle 0|\underset{\sim}{e}^2|0\rangle = \frac{\hbar ck}{2\varepsilon_o V} \tag{2.23}$$

The associated fluctuations can be seen as causing spontaneous emission from excited atoms and molecules, and the Lamb shift.

3. The radiation-molecule coupling

In applications to intermolecular interactions, the term H_{int} in the complete Hamiltonian (2.1) includes the coupling of the molecules to the radiation field plus the electrostatic coupling of molecules to each other. To be specific let us take the case of two hydrogen atoms A and B, with nuclei a and b and electrons 1 and 2. The molecular coupling term is V_{AB},

$$V_{AB} = \frac{1}{R_{ab}} + \frac{1}{r_{12}} - \frac{1}{r_{a2}} - \frac{1}{r_{b1}} \tag{3.1}$$

which, in dipole approximation, becomes

$$V_{AB} \sim \frac{\mu_i(1)\ \mu_j(2)}{4\pi\varepsilon_o R^3}\ \beta_{ij} \tag{3.2}$$

The molecule-radiation coupling part of H_{int} is, in the minimal coupling picture,

$$H_{rad\ int} = \frac{e}{m}\left\{\underset{\sim}{p}\cdot\underset{\sim}{a}(\underset{\sim}{R}_A) + \underset{\sim}{p}\cdot\underset{\sim}{a}(\underset{\sim}{R}_B)\right\} + \frac{e^2}{2m}\left\{\underset{\sim}{a}^2(\underset{\sim}{R}_A) + \underset{\sim}{a}^2(\underset{\sim}{R}_B)\right\} \tag{3.3}$$

where m is the electron mass, and $\underset{\sim}{p}$ is the canonical momentum, not simply the kinetic momentum,

$$\underset{\sim}{p} = \underset{\sim}{p}^{kin} - e\underset{\sim}{a}(\underset{\sim}{q}) \tag{3.4}$$

which is conjugate to the displacement $\underset{\sim}{q}$ in the presence of a radiation field of vector potential $\underset{\sim}{a}$. -e is the electron charge.

Thus the total interaction (3.1) plus (3.3) has an electrostatic part, implying that the effect of charge displacements at A is felt instantaneously at B. It also has a part (3.3) in which the influence of A is conveyed radiatively to B, i.e. at the speed of light. It is of course obvious that the total interaction must be fully retarded; there can be no uncancelled instantaneous part. The explanation is that in Coulomb gauge $\underset{\sim}{a}$ contains an unretarded component, the effect of which through (3.3) exactly cancels (3.2) leaving a fully retarded result. This holds in general.

An alternative and often more convenient form of coupling operator is the multipolar interaction. In the dipole approximation the momentum conjugate to $\underset{\sim}{q}$ is purely kinetic, but the vector $\underset{\sim}{\Pi}(\underset{\sim}{r})$ conjugate to $\underset{\sim}{a}$ is not simply the (transverse) electric field as in (2.15) but is the negative of the transverse part of the microscopic displacement $\underset{\sim}{d}(r)$,

$$\Pi(\mathbf{r}) = - \mathbf{d}^{\perp}(\mathbf{r}) \tag{3.5}$$

where

$$\mathbf{d}(\mathbf{r}) = \varepsilon_o \mathbf{e}(\mathbf{r}) + \mathbf{p}(\mathbf{r}) \tag{3.6}$$

and, in dipole approximation, the polarization is given by

$$\mathbf{p}(\mathbf{r}) = \boldsymbol{\mu}(1)\ \delta\left(\mathbf{r} - \mathbf{R}_A\right) + \boldsymbol{\mu}(2)\ \delta\left(\mathbf{r} - \mathbf{R}_B\right) \tag{3.7}$$

where the $\boldsymbol{\mu}$ are dipole moment operators and $\mathbf{R}_A$ and $\mathbf{R}_B$ the positions of the atoms. For the applications in this article the complete interaction for the two electron case is

$$H_{int} = - \varepsilon_o^{-1}\boldsymbol{\mu}(1)\cdot\mathbf{d}^{\perp}(\mathbf{R}_A) - \varepsilon_o^{-1}\ \boldsymbol{\mu}(2)\cdot\mathbf{d}^{\perp}(\mathbf{R}_B) \tag{3.8}$$

and for the radiation Hamiltonian,

$$H_{rad} = \frac{1}{2} \int \left\{ \frac{\mathbf{d}^{\perp 2}}{\varepsilon_o} + \varepsilon_o c^2 \mathbf{b}^2 \right\} d^3\mathbf{r} \tag{3.9}$$

4. The resonance coupling

The applications of the method will be shown in key examples. The first is the resonance intermolecular interaction, which is the simplest, and in which the primitive coupling types appear in a calculation of second order in the interaction (3.8).

Given identical atoms A and B, one in an excited level, the pair states $|E_n^A, E_o^B>$ and $|E_o^A, E_n^B>$ are of equal energy $E_n + E_o$. Let us suppose that $|E_n> \leftarrow |E_o>$ is dipole allowed. From a stationary state viewpoint at small separations the degeneracy is split, giving states

$$\Psi_{\pm} = 2^{-1/2}\left\{\left|E_n^A,\ E_o^B\right\rangle \pm \left|E_o^A,\ E_n^B\right\rangle\right\} \tag{4.1}$$

with energy separation $\Delta E \simeq 2|M|$, where

$$M = \left\langle E_n^B, E_o^A \mid H_{int} \mid E_n^A, E_o^B \right\rangle \quad (4.2)$$

In the case that the transition dipole moments are parallel in A and B only ψ_+ in (4.1) is allowed in transitions from $|E_o^A, E_o^B\rangle$, with moment

$$\underset{\sim}{\mu}_+ = 2^{-1/2}\left\{\underset{\sim}{\mu}^{on}(A) + \underset{\sim}{\mu}^{on}(B)\right\}$$

$$= 2^{1/2}\underset{\sim}{\mu}^{on} \quad (4.3)$$

where $\underset{\sim}{\mu}^{on}$ is the moment common to A and B. Thus the transition rate for transitions ending on ψ_+ is twice that for the separate molecules. From a dynamical viewpoint, with the system in the state $|E_n^A, E_o^B\rangle$ at zero time, the excitation oscillates between A and B with frequency $|M|/\hbar$ s^{-1}; the probability that the excitation is on A or on B depends on the time sinusoidally.

Although resonance coupling is the simplest of the intermolecular types, its application to particular systems needs careful analysis, since the separate excited molecules, or the coupled pair, can decay by spontaneous emission to the ground state by photon emission. The matrix element for emission is given in (4.4),

$$\left\langle 1(\underset{\sim}{k},\lambda);\ E_o|H_{int}|\ E_n;\ 0\right\rangle$$

$$= \left\langle 1(\underset{\sim}{k},\lambda);\ E_o|\ -\ \varepsilon_o^{-1}\underset{\sim}{\mu}.\underset{\sim}{d}^{\perp}\ |E_n;\ 0\right\rangle \quad (4.4)$$

The initial state has the molecule in its n-th level and no photons in the radiation field (vacuum field). The fluctuations in the electric field (2.23) in this vacuum state cause transitions to the final state, consisting of ground state molecule and 1 photon of the resonant mode in the field.

The radiative lifetime for a typical molecular transition in the near UV is 10^{-9} - 10^{-8} s. Thus whether a stationary state picture applies, or a dynamical picture involving the transfer rate, depends on the strength of intermolecular coupling. At short distances, with strong coupling, a stationary state of the molecule-pair is formed. At long distances radiative decay occurs before photon exchange between the molecules. The crossover in behaviour occurs for $|M|$ in the order of magnitude 10^{-2} cm^{-1}.

Molecular quantum electrodynamics deals with resonance interaction as the coupling of the initial $|i\rangle$ and final $|f\rangle$ states through intermediate states I and I',

$$\left.\begin{aligned} |i\rangle &= |E_n^A,\ E_o^B;\ 0\rangle \\ \\ |f\rangle &= |E_o^A,\ E_n^B;\ 0\rangle \end{aligned}\right\} \qquad (4.5))$$

$$\left.\begin{aligned} |I\rangle &= |E_o^A,\ E_o^B;\ 1(\underset{\sim}{p},\lambda)\rangle \\ \\ |I'\rangle &= |E_n^A,\ E_n^B;\ 1(\underset{\sim}{p},\lambda)\rangle \end{aligned}\right\} \qquad (4.6)$$

In (4.5) the zeros denote the vacuum state of the radiation; in (4.6) the intermediate states have 1 photon of mode $(\underset{\sim}{p},\lambda)$; and the two molecules either both in the ground state, or both in the n-th excited states. The intermediate states are virtual, with no energy conservation; they are summed over all values of the wavevector $\underset{\sim}{p}$ and both of the polarizations λ. The matrix elements can be evaluated conveniently with use of time ordered diagrams, which for the resonance problem are shown in Figure 1.

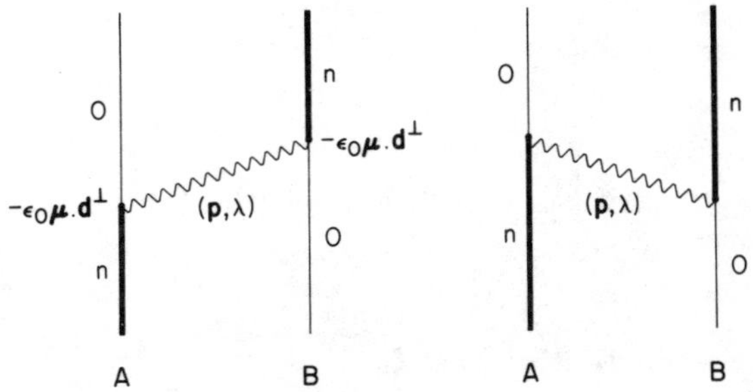

Figure 1. Time-ordered diagrams for the resonance coupling of identical molecules.

Time moves upward. At each vertex (junction of wavy and straight lines) there is a change of state caused by the vertex interaction:

$$- \varepsilon_o^{-1} \underset{\sim}{\mu}(A/B).\underset{\sim}{d}^{\perp}(\underset{\sim}{R}_{A/B}) \tag{4.7}$$

which causes a transition to the intermediate state |I> (left hand diagram) or |I'> (right hand diagram). At the second vertex

$$- \varepsilon_o^{-1} \underset{\sim}{\mu}(A/B).\underset{\sim}{d}^{\perp}(\underset{\sim}{R}_{B/A}) \tag{4.8}$$

causes a transition from the intermediate to the final state. The sum of the matrix elements gives the result for the fully retarded matrix element (4.9) [4],

$$M = \mu_i^{on}(A)\ \mu_j^{no}(B) \left\{ V_{ij}(k,\underset{\sim}{R}) + iW_{ij}(k,\underset{\sim}{R}) \right\} \tag{4.9}$$

where $V_{ij}(k,\underset{\sim}{R})$ is the real part of the interaction potential

$$V_{ij}(k,\underset{\sim}{R}) = \frac{1}{4\pi\varepsilon_o R^3} \left[\beta_{ij} \left(\cos kR + kR\sin kR \right) - \right.$$

$$\left. - \alpha_{ij} \left(k^2R^2 \cos kR \right) \right] \tag{4.10a}$$

and $W_{ij}(k,\underset{\sim}{R})$ the imaginary part,

$$W_{ij}(k,\underset{\sim}{R}) = \frac{1}{4\pi\varepsilon_o R^3} \left[\beta_{ij} \left(- \sin kR + kR\cos kR \right) + \right.$$

$$\left. + \alpha_{ij} \left(k^2R^2 \sin kR \right) \right] \tag{4.10b}$$

k is the resonant wave number, $E_n - E_o = \hbar ck$.

β_{ij} is the dyadic for full dipolar interaction introduced by (1.1). α_{ij} is the transverse dipole coupling dyadic

$$\alpha_{ij} = \delta_{ij} - \hat{R}_i \hat{R}_j \tag{4.11}$$

giving terms in the potential (4.10) only from the components of the dipoles transverse to their join $\underset{\sim}{R}$, with zero from the longitudinal components. This is characteristic of radiative coupling, effected through the electric field $\underset{\sim}{e}$ transverse to the propagation direction. At distances short compared with the transition wavelength, $kR\underset{\sim}{-}0$, the potential (4.10) is of the Coulombs law form

$$V_{ij}(0,R) = \frac{1}{4\pi\varepsilon_o R^3}\,\beta_{ij} \tag{4.12}$$

At long distances, the final terms dominate

$$V_{ij}(k,\underset{\sim}{R}) \sim -\frac{k^2}{4\pi\varepsilon_o R}\cos kR\;\alpha_{ij} \tag{4.13a}$$

and

$$W_{ij}(k,\underset{\sim}{R}) \sim \frac{k^2}{4\pi\varepsilon_o R}\sin kR\;\alpha_{ij} \tag{4.13b}$$

kR is the number of reduced wavelengths $\lambda = k^{-1}$ in the separation R. Let us estimate the distance at which the rate of exchange of excitation becomes too slow to compete with radiative decay. If the radiative decay lifetime is 3×10^{-3} cm^{-1} the excitation exchange frequency, at 3×10^{-8} s^{-1} is such that the probability of excitation remaining in the two-molecule system has dropped to 1/2.7 in the period of one complete cycle of excitation exchange. One can take this as a cut-off point; beyond it excitation exchange is slow compared with spontaneous emission. For a representative molecular transition (unit oscillator strength, wave number 20×10^3 cm^{-1}) the static contribution (4.10) is about 10^{-3} cm^{-1} at a separation of 35 nm, or ~ 0.11 reduced wavelengths. At this separation the wave-zone contribution (4.13) is ~ 1.2% of the near zone (4.12), and can have physical significance as the separation approaches the decay cut-off. At such intermediate distances the wave-zone coupling partially cancels the transverse part of the static coupling, and at long distances gives the sole contribution. The terms in the potential (4.10) with overall R^{-2} dependence, such as

$$\frac{1}{4\pi\varepsilon_o R^3}\,kR\sin kR\;\beta_{ij} \tag{4.14}$$

are in phase quadrature with the static and wave-zone terms, and under the conditions just given have already changed the static contribution by 10%. Its importance in modulating the static term cannot be ignored.

5. Magnetic moment contributions to resonance interaction

Chiral molecules, defined by lack of any improper rotation axis, and so limited to point groups C_1, C_n, D_n, T, O and I, $n \geq 2$, possess some optical transitions that are allowed both to electric and magnetic dipole radiation. Again the simplest application is to resonance coupling, which illustrates features that appear in the higher order chiroptical properties. The operator H_{int} in the complete Hamiltonian (3.7) now is supplemented by terms for the magnetic coupling,

$$H_{int} = - \sum_{J=A,B} \left\{ \varepsilon_o^{-1} \underset{\sim}{\mu}(J).\underset{\sim}{d}^{\perp}(\underset{\sim}{R}_J) - \underset{\sim}{m}(J).\underset{\sim}{b}(\underset{\sim}{R}_J) \right\} \tag{5.1}$$

where $\underset{\sim}{m}(J)$ is the magnetic moment operator for molecule J.

The time-ordered graphs in Fig. 1 are supplemented by similar graphs where either both vertices become $-\underset{\sim}{m}.\underset{\sim}{b}$, or there is one each of $-\varepsilon_o^{-1}\underset{\sim}{\mu}.\underset{\sim}{d}$ and $-\underset{\sim}{m}.\underset{\sim}{b}$. The first type, for the magnetic-magnetic interaction, is not of interest.

The second mixed electric-magnetic type contributes extra terms to the matrix element, to be added to the electric-electric term (4.9),

$$M' = \text{Im}\left\{\mu_i^{on}(A)m_j^{no}(B) + m_j^{no}(A)\mu_i^{on}(B)\right\} \times$$

$$\times \left\{U_{ij}(k,\underset{\sim}{R}) + iX_{ij}(k,\underset{\sim}{R})\right\} \tag{5.2}$$

where U_{ij}, analogous to V_{ij} in (4.10), is

$$U_{ij}(k,\underset{\sim}{R}) = \frac{1}{4\pi\varepsilon_o c}\frac{\hat{R}_k}{R^3}\left\{kR\cos kR + k^2R^2\sin kR\right\}\varepsilon_{ijk} \tag{5.3a}$$

and

$$X_{ij}(k,\underset{\sim}{R}) = \frac{1}{4\pi\varepsilon_o c}\frac{\hat{R}_k}{R^3}\left\{-kR\sin kR + k^2R^2\cos kR\right\}\varepsilon_{ijk} \tag{5.3b}$$

where the tensor properties are in the Levi-Civita tensor ε_{ijk}, $\varepsilon_{ijk} = \pm 1$ for cyclic and anti-cyclic index order, and $= 0$ for repeated indices. (5.3) does not have a nonzero static limit at $kR = 0$, in line with the fact that static electric and magnetic moments do not couple. The near zone behaviour for small kR is

$$U_{ij} \sim \frac{1}{4\pi\varepsilon_o c}\frac{k\hat{R}_k}{R^2}\varepsilon_{ijk} \tag{5.4}$$

In direction dependence (5.2) is greatest when the (i,j,k) form a right-angled triad, as when the electric and magnetic transition moments are at right-angles to each other and also to the propagation direction of the radiation.

The electric-magnetic term is much smaller than the electric-electric. In the first place the magnetic dipole matrix elements m_j^{no} are typically at least 2 orders of magnitude less than the μ_j^{no}; and secondly the leading distance term $\sim kR^{-2}$ is 2 or 3 orders down from R^{-3} in the near zone. However that is not to say that magnetic terms are unimportant. In differential effects[5,6], where the large electric-electric terms cancel, such as chiral discrimination, they are easily measurable. Also in radiation-molecule phenomena, not discussed here, they can be dominant, for example, in optical rotation and circular dichroism.

From another point of view we can use the correspondence principle to rationalize the quantum electrodynamical expressions (4.9) and (5.2). We would then treat the coupling of classical oscillating dipoles via the classical Maxwell field. The spatial part of the (complex) electric field $\underset{\sim}{E}(\underset{\sim}{R}_B)$ due to the dipole at $\underset{\sim}{R}_A$ is [7],

$$\underset{\sim}{E}(\underset{\sim}{R}_B) = \frac{e^{-ikR}}{4\pi\varepsilon_o}\left[\left(\frac{1}{R^3} + \frac{ik}{R^2}\right)\left\{3\hat{\underset{\sim}{R}}(\hat{\underset{\sim}{R}}.\underset{\sim}{\mu}(A)) - \underset{\sim}{\mu}(A)\right\} - \frac{k^2}{R}\left\{\hat{\underset{\sim}{R}} \times (\hat{\underset{\sim}{R}} \times \underset{\sim}{\mu}(A))\right\}\right] \tag{5.5}$$

where $\underset{\sim}{R} = \underset{\sim}{R}_B - \underset{\sim}{R}_A$.

The coupling to a dipole $\underset{\sim}{\mu}(B)$ is, expressed in components, and left in complex form,

$$- \underset{\sim}{E}(\underset{\sim}{R}_B).\underset{\sim}{\mu}(B) = \frac{\mu_i(A)\mu_j(B)}{4\pi\varepsilon_o} \cdot \left\{\left(\frac{1}{R^3} + \frac{ik}{R^2}\right)\beta_{ij} - \frac{k^2}{R}\alpha_{ij}\right\}e^{-ikR} \tag{5.6}$$

which can be compared with the expression (4.10) from quantum electrodynamics. In classical electrodynamics the R^{-3}, R^{-2} and R^{-1} terms are described as 'static', 'induction' and 'radiation'.

In the classical theory of the Maxwell field, net outward energy flow from an oscillating dipole, calculated from the Poynting vector, is contributed only by the radiation term. Energy flow from the other terms oscillates between inward and outward, with zero average. This corresponds in the quantum theory to radiative decay of an excited system by energy loss into modes of the photon field. Conservation of energy holds only when the system is taken to be composed of molecule plus radiation.

The magnetic field due to the electric dipole at $\underset{\sim}{R}_A$ is

$$\underset{\sim}{B}(\underset{\sim}{R}_A) = - \frac{ik e^{-ikR}}{4\pi\varepsilon_o c}\left(\frac{1}{R^2} - \frac{ik}{R}\right) \underset{\sim}{\mu}(A) \times \underset{\sim}{R} \tag{5.7}$$

Both the first (induction) and second (radiation) terms in the magnetic field (5.7) are purely transverse. There is no longitudinal component, in line with remarks made following equation (5.4). The electric dipole-driven magnetic field vanishes for zero frequency, and in the near zone limit of $kR = 0$.

6. Effect of radiation on intermolecular coupling

The next application of quantum electrodynamical methods is to calculate the effect of intense radiation on the interaction of a pair of molecules. This exemplifies cases in which the influence of retardation, and specifically wave zone terms are dominant.

The dispersion interaction of two atoms or molecules depends on distance as R^{-6} at short distances and on R^{-7} at long distances. If the coupled molecules are strongly irradiated say by a laser beam, there are additional terms going as R^{-1}. This interesting result, first found by Thirunamachandran [8,9], suggests that although individual pairwise radiation-induced interactions are small the total coupling of a chosen molecule to its neighbours will be the result of many individual couplings to molecules within the range of the slowly decreasing interaction. Energy shifts of at least a few cm^{-1} are possible under currently available laser powers from this source. This is an example in which the effects of radiation, over and above the van der Waals energy, are entirely dynamic. The electrostatic limit corresponding to a zero frequency transverse field, is zero. We confirm this point first by taking a static field $\underset{\sim}{E}$ applied to molecules with isotropic polarizability and in the gas or liquid phase. Each molecule will have an induced moment $\underset{\sim}{\mu} = \alpha\underset{\sim}{E}$. Since the molecules are uniformly distributed, each will be surrounded by a spherical distribution of uniform polarization, giving zero electric field at its centre. There is zero energy shift.

However for a radiation field of irradiance I the energy shift in the interaction of two molecules separated by $\underset{\sim}{R}$ is found to be given by (6.1),

$$\Delta E(\underset{\sim}{R}) = \frac{I}{\varepsilon_o c}\, \bar{e}_i^{(\lambda)}(\underset{\sim}{k})\, e_j^{(\lambda)}(\underset{\sim}{k})\alpha^A(k)\alpha^B(k) \times$$

$$\times V_{ij}(k\underset{\sim}{R}) \cos(\underset{\sim}{k}.\underset{\sim}{R}) \tag{6.1}$$

the e_i are polarizations, $\alpha^A(k)$ is the dynamic polarizability at frequency ck, and $V_{ij}(k,\underset{\sim}{R})$ is the potential (4.10). Again we must take account of the uniform molecular distribution, in which each direction of R is equally probable. The appropriate average of (6.1) gives

$$\Delta E = -\frac{I}{8\pi\varepsilon_o^2 c}\cdot\frac{\alpha^A(k)\alpha^B(k)}{R^3}\left(kR\sin 2kR + 2\cos 2kR - \frac{5\sin 2kR}{kR} - \frac{6\cos 2kR}{k^2R^2} + \frac{3\sin 2kR}{k^3R^3}\right) \tag{6.2}$$

which, in the near zone reduces to an attractive interaction going as R^{-1},

$$\Delta E \sim -\left(\frac{11\ I\ k^2}{60\pi\varepsilon_o^2 c}\right)\frac{\alpha^A(k)\alpha^B(k)}{R} \tag{6.3}$$

In the averaging leading to (6.2) the terms in V_{ij} depending on β_{ij} vanish as explained for a static electric field. The surviving terms are for the radiative coupling of the induced moments, namely the coupling of the moments transverse to $\underset{\sim}{R}$. These do not cancel over spherical sums, and give the characteristic R^{-1} dependence.

The result is a small energy shift, proportional to the irradiance of the laser beam. If the beam frequency is chosen to be near-resonant with a molecular transition frequency the effect can be greatly enhanced.

7. Induced circular dichroism

The final example is typical of cases where the important properties have their origin in the near-zone, but where quantum electrodynamics enables a physically transparent approach. This is because the radiation-molecule and molecule-molecule couplings are treated from a single point of view.

The system consists of two molecules A and C, the first achiral and the second chiral, together with a radiation field from which photons are to be absorbed by the combined A-C system. Because of the coupling between A and C, chirality is induced in A. The result is that A becomes optically active. This activity is weak, and its detection difficult. Experimentally a successful method is the measurement of induced circular dichroism. It is a measurement of circular dichroism at an absorption frequency of the achiral molecule. Without an environment of chiral molecules there is no circular dichroism at any frequency. This observation of circular dichroism is diagnostic for the induction of optical activity in an achiral molecule by a chiral one. The first experimental measurements (in the n-π* transition of aromatic ketones) [10] have been followed by large numbers of others.

If the electric dipole allowed transition in the achiral molecule is of frequency ck, there is the underlying process of ordinary absorption at this frequency, showing no circular dichroism. This is illustrated in the left-hand time-ordered diagram in Fig 2,

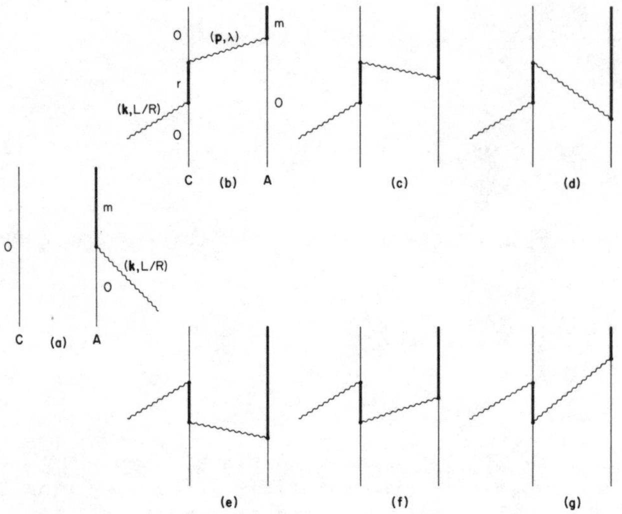

Figure 2. Time-ordered diagrams for induced circular dichroism. Left hand: first order absorption process, unaffected by C. Right hand: third order process.

The interaction Hamiltonian, analogous to (5.1) is

$$H_{int} = -\varepsilon_0^{-1}\mathbf{\mu}(A).\mathbf{d}^{\perp}(\mathbf{R}_A) - \varepsilon_0^{-1}\mathbf{\mu}(C).\mathbf{d}^{\perp}(\mathbf{R}_C) - \mathbf{m}(C).\mathbf{b}(\mathbf{R}_C) \quad (7.1)$$

Since A is achiral there is no magnetically active dipole transition. For C both electric and magnetic terms must appear. The transition matrix element for the process of independent absorption (left hand diagram, Fig. 2), is first order,

$$M_{fi}^{(1)(L/R)} = -i\left(\frac{n\hbar ck}{2\varepsilon_0 V}\right)^{1/2} e_i^{(L/R)}(\mathbf{k})\mu_i^{mo} \quad (7.2)$$

where n is the number of photons in the incident beam in the $\mathbf{k}$ mode, and (L/R) indicates left and right circular polarization. The transition rates for the two polarizations are equal, there being no chirality in A to this order.

There are no terms of second order. The third order terms also give equal L and R rates, but in conjunction with the first order (7.2) there is a cross-term (interference term) which discriminates between the rates for the two polarizations. The total matrix element [11] to third order for absorption of a ($\mathbf{k}$,L/R) photon is

$$M_{fi}^{(L/R)} = M_{fi}^{(1)(L/R)} + M_{fi}^{(3)(L/R)}$$

$$= - i \left(\frac{n\hbar ck}{2\varepsilon_o V}\right)^{1/2} e_i^{(L/R)}(\underset{\sim}{k})\mu_k^{mo} \Bigg[\delta_{ik} -$$

$$- \left\{\left(\alpha_{ij} \mp (i/c)G_{ji}\right)V_{jk} + iG_{ij}U_{jk}\right\}e^{i\underset{\sim}{k}.\underset{\sim}{R}}\Bigg] \qquad (7.3)$$

there

$$G_{ij} = \sum_r \left\{\frac{\mu_i^{or} m_j^{ro}}{E_{ro} - \hbar ck} + \frac{m_j^{or} \mu_i^{ro}}{E_{ro} + \hbar ck}\right\} \qquad (7.4)$$

G_{ij} is the dynamic electric-magnetic polarizability analogous to the electric-electric α_{ij}.

The differential rate of left/right absorption for on A-C pair freely rotating is

$$\left\langle\Gamma^{(L)} - \Gamma^{(R)}\right\rangle = \frac{8 \mathcal{I} k^2}{135\pi\hbar^2 \varepsilon_o^2 c} |\mu^{mo}|^2 \left(\mathrm{Im}G_{\nu\nu}\right)\frac{1}{R} \qquad (7.5)$$

for pairs for which kR is small. $\mathcal{I}$ is the radiant energy density.

(7.5) is the so-called dynamic contribution to induced circular dichroism. There are contributions from other sources, particularly from permanent electric moments which are expected to be important in molecule pairs in orientations that are fixed relative to the intermolecular join.

These methods are easily adaptable to a problem closely related to induced CD. The intermolecular coupling that induces chiral properties in the achiral molecule A also induces a change in the chiral molecule C. One effect is that its optical activity is changed. Given that the chiroptical properties depend on Im $\underset{\sim}{\mu}^{om}.\underset{\sim}{m}^{mo}$ for the various transitions of C, we can expect that by coupling to A the effective value of this quantity will be changed. A part of the electric transition moments for transitions in A will be transferred to C, i.e. 'stolen' by C analogously to vibronic intensity stealing in forbidden molecular electronic transitions.

The translation of this picturesque physical description into a quantum electrodynamical scheme, dealing in the same framework with the intermolecular and radiation effects, is again via the diagrammatic technique [12]. The absorbing entity is C instead of A. If we interchange the labels A and C in the diagrams of Figure 2, we have diagrams for the effect of A on the transition of C that shows circular dichroism. The left hand diagram includes no effect of A. This first order matrix element has to be added to third order elements from the right hand diagrams and others like it, corresponding to different time orderings. The differential effect of A on the L and R absorption rates of C arises from interference between the first and third order parts. In this case the result is a fractionally small modification of the properties of C, whereas in induced CD it is responsible for the entire effect.

To illustrate the quantum electrodynamical result we give the correction terms to the natural circular dichroism of C. The relative orientations of A and C are fixed; the orientations of the molecular join $\underset{\sim}{R}$ are randomly averaged relative to the direction $\hat{\underset{\sim}{k}}$ of the light beam

$$\frac{Ii}{2\varepsilon_o\hbar^2c^2}\Bigg[m_k^{mo}(C)\mu_l^{mo}(C)\alpha_{mn}^{A}(k)V_{lm}(k,\underset{\sim}{R})$$
$$\times\left\{\left(\delta_{kn}-\hat{R}_k\hat{R}_n\right)\frac{\sin kR}{kR}+\left(\delta_{kn}-3\hat{R}_k\hat{R}_n\right)\left(\frac{\cos kR}{k^2R^2}-\frac{\sin kR}{k^3R^3}\right)\right\}-$$
$$-c\varepsilon_{knp}\mu_k^{mo}(C)m_l^{mo}(C)\alpha_{mn}^{A}(k)U_{lm}(k,\underset{\sim}{R})\left\{\frac{\cos kR}{kR}-\frac{\sin kR}{k^2R^2}\right\}\hat{R}_p\Bigg] \quad (7.6)$$

In the near-zone the random average over relative molecular orientations is zero. Detection however seems possible in a mixed crystal.

8. References

[1] Casimir, H.B.G. and D. Polder, 'Influence of retardation of the electrostatic interaction between atoms in the London-van der Waals forces' Phys. Rev. (1948) **73**:360.

[2] Tabor, D., and R.H.S. Winterton, 'The direct measurement of normal and retarded van der Waals forces' Proc. Roy. Soc. Lond. (1969) **A312**:435.

[3] Craig, D.P. and T. Thirunamachandran, Molecular Quantum Electrodynamics, Academic Press, London, 1984.

[4] Power, E.A., and S. Zienau, 'Coulomb gauge in non-relativistic quantum electrodynamics and the shape of spectral lines' Phil. Trans. Roy. Soc. Lond. (1959) **A251**:54.

[5] Craig, D.P. and D.P. Mellor,
'Discriminating interactions between chiral molecules'
Topics in Current Chemistry (1976) **63**:1.

[6] Craig, D.P., E.A. Power and T. Thirunamachandran,
'The interaction of optically active molecules'
Proc. Roy. Soc. Lond. (1971) **A322**:165.

[7] Stratton J.A.,
Electromagnetic Theory, McGraw Hill, New York, 1941.

[8] Thirunamachandran, T.,
'Intermolecular interactions in the presence of an intense radiation field'
Mol. Phys. (1980) **40**:393.

[9] Taylor, M.D. and T. Thirunamachandran,
'Discriminatory interactions between chiral molecules in the presence of an intense radiation field'
Mol. Phys. (1983) **49**:881.

[10] Bosnich, B.,
'Asymmetric syntheses, asymmetric transformations, and asymmetric inductions in an optically active solvent'
J. Amer. Chem. Soc., (1967) **89**:6143.

[11] Craig, D.P., E.A. Power and T. Thirunamachandran,
'The dynamic terms in induced circular dichroism'
Proc. Roy. Soc. Lond. (1976) **A348**:19.

[12] Craig, D.P., and T. Thirunamachandran,
'Change in chiroptical properties caused by intermolecular coupling'
Proc. Roy. Soc. Lond. (1987) **A410**:337.

INTRAMOLECULAR DYNAMICS

R.D. LEVINE
The Fritz Haber Research Center for Molecular Dynamics
The Hebrew University
Jerusalem 91904, Israel
and
Department of Chemistry
Harvard University
Cambridge, MA 02138, USA

ABSTRACT. There is currently both experimental and theoretical evidence that also in systems of many interacting atoms there can be dynamical selectivity with respect to initial mode of excitation and specificity with respect to final outcomes. This lecture provides a brief outline of the newly available results and examines in detail the dynamics of highly vibrationally excited acetylene.

1. Introduction

The conventional wisdom for intramolecular dynamics is that on time scales of chemical interest, highly excited states of polyatomic molecules behave statistically. In other words, there is enough time for the excess energy to be shared amongst all modes. The 'chemistry' of the molecule will then be governed only by its total energy rather than by any detail of its initial preparation. The earlier evidence in support of this point of view [1], provided mainly by chemical activation experiments [2], has more recently been augmented using activation via optical excitation and in particular multiple (infrared) photon absorption [3,4]. The purpose of this lecture is to go beyond this point of view. We point out that there is currently a wealth of evidence, for many different types of systems, that the conventional wisdom needs refinement. We then examine a particular example in some detail.

Lecture at the 6'th International Congress of Quantum Chemistry

J. Jortner and B. Pullman (eds.), Perspectives in Quantum Chemistry, 41–55.

2. Background

The need for reexamination of intramolecular dynamics is stimulated by the experimental progress which has resulted in both shorter times in which systems can be excited and/or probed and in systems where the inherent time scales for energy redistribution are longer than for ordinary polyatomic molecules. Starting with the latter, experiments are now possible on systems of many interacting atoms where the range of vibrational frequencies is far wider than is the case for polyatomic molecules made up from atoms of comparable mass. This is achieved in primarily two ways. The first is to have forces of qualitatively different strengths coupling the atoms in the system. An example is a van der Waals adduct of two molecules. The chemically bonded atoms interact much more strongly than those coupled by the (nominally, non-bonding) van der Waals forces. Clusters, molecules physisorbed on surfaces and chemical reactions in liquids are additional examples.

The intramolecular dynamics of systems of atoms which interact with forces of different scales is so rich that we discuss some examples in more detail in the next section.

The other way to introduce a significant frequency mismatch is by changes in the mass. (Of course, one can do both, as in a chemical reaction in solution where the reactants and solvent molecules have quite different masses). Even in chemical activation days, the C-H modes presented a problem and our example of acetylene is similar in that respect. An opposite extreme, also encountered first in chemical activation, is heavy atom blocking, [5].

The tighter preparation of the initial state and its probing can be achieved in either the time or the frequency domains. The experimental ability to observe the system during the chemical act and hence to monitor the dynamics in real time [6] has been widely acclaimed. It should however be emphasized that one can get equivalent information from the frequency response, whether it is the emitted fluorescence [7,8] or the absorption spectrum [9,10] of the system [1]. The work of Heller [11,12] on the equivalence of short time dynamics and coarse grained spectra (and *vice versa*) has contributed significantly to this point.

Since comparatively short propagation times are required to generate the major spectral features, there is a renaissance of time-dependent quantal computational methods. These include exact [13] and approximate [14,15] wave packet propagation, time-dependent self-consistent-field theories [16], path integral techniques [17], algebraic-based approaches [18] and RRGM-based methods [19]. New methods for the analysis of complex spectra, which go beyond the near neighbors spacing distribution [20], are also being developed [21-24].

Progress in non-linear dynamics has also had a significant impact. The importance of 'resonance' frequency match or lack thereof contributed to the development of local mode models [25-27]. Another central idea is that of Cantori [28,29] (closely related to vague tori [30] and to intermittency). This is the observation that when two modes of evolution are energetically accessible, the system will not necessarily smoothly change between the two. Rather, it will spend quite some time in one mode (as if it were the only allowed mode) and then make a rather sudden, impulsive transition to the other. The sudden onsets of new modes of motion will be quite evident in the example of acetylene, as discussed in detail below. It is because of this characteristic that we are able to delineate distinct stages in the temporal evolution of highly excited acetylene.

Furthermore, because these stages correspond to different time scales, they have their signature in the observed spectral features.

It is this separation of time scales with a corresponding (yet tentative) experimental identification via different levels of spectral resolution that make acetylene a particularly interst-ing test case. Other molecules whose intramolecular dynamics have been theoretically studied in detail include C_6H_6 [31-36], CH_2O [37,38] and H_3^+ [39-42]. So far, we have discussed the intramolecular dynamics on the ground electronic surface. There is, of course, a wealth of data and theory for the dynamics on the first electronically excited surface. We refer to recent reviews for more details [43,44].

3. Reactive Processes

Most of the systems mentioned so far were bound states. In this section we briefly review theoretical studies of reactive processes in many atom systems. In each case one can regard the collection of interacting atoms as a polyatomic molecule. Simple unimolecular rate theory will then suggest that any selective initial excitation will rapidly equilibrate. Yet all three classes of processes to be discussed exhibit distinct dynamical features.

3.1 VAN DER WAALS MOLECULES AND CLUSTERS

Deviations from the simple RRKM picture are to be expected for molecules where there is considerable frequency mismatch among the coupled coordinates [45]. This has been clearly demonstrated for van der Waals molecules [45-47] and, in particular for the $I_2{\cdot}He$ complexes [47-51]. What is of particular interest here is the mechanism of dissociation. The trajectory simulations show that the motion in the van der Waals well is largely unperturbed (due to the poor resonance between its frequency and that of I_2) until a rare short range repulsive interaction between He and an I atom which delivers an impulse sufficient to dissociate the van der Waals bond. The important point is that the rate of dissociation is essentially the rate of energy transfer [52]. This is not found to be the case for van der Waals systems (e.g., Ar_3 [53] where the frequencies are commensurate, energy exchange is facile and the RRKM point of view is quantitatively useful [54]. Even for rare gas clusters [55,56] e.g., Br_2 in Ar and for collisions such as Xe + Ar_2 [57] one finds that impulsive collisions provide the mechanism for energy transfer and evaporation. A similar bottleneck is operative when a vibrationally excited physisorbed diatomic molecule, with enough internal excitation to break the van der Waals bond to the surface, desorbs [58].

3.2 SURFACE DESORPTION

The desorption from a heated surface [59,60] can be discussed from a transition state theory point of view. For the purpose of considering the competition between desorption and surface reaction [61,62] it is convenient to regard the adsorbed species and the solid together as a polyatomic molecule that comes to thermal equilibrium more rapidly than any reactive change. The observed preferential desorption at higher heating rates is then due to the desorption rate constant having both a higher pre-exponential factor and a higher activation energy. At low heating rates, the temperature of the polyatomic system rises slowly and surface reaction dominates. For a rapid rise in the temperature, the temperature becomes rapidly high enough that the higher pre-exponential factor makes the desorption rate faster. It has been suggested [63] that this 'equipartition' picture may fail for physisorbed molecules. The argument is a continuation of the one in section 3.1. There, due to frequency mismatch there is no facile energy transfer from the chemical to the van der Waals bond. Here, due to frequency mismatch, energy does not readily flow from the low frequency physisorption bond to the internal modes of the physisorbed species. The physisorption bond acts as a bottleneck to energy flow from the solid to the molecule. Some support for this notion is provided by experiments [64,65] that suggest that even large polyatomics tend to desorb internally rather cold upon laser induced heating of the surface. Classical molecular dynamics simulations of desorption of diatomics from rapidly heated clusters [66] also show that physisorbed diatomics are desorbed cold. Even more important, the simulations show that desorption follows immediately an impulsive energy transfer from a cluster atom. As a further check, classical simulations [67] for diatomics tightly bound to the cluster or for diatomics which themselves are weakly bound (so that their internal frequency is commensurate with that of the physisorption bond or of the cluster), behave in a much more RRKM-like fashion, for the same heating rates that lead to selective desorption in the previous case.

3.3 REACTIONS IN LIQUIDS

Molecular dynamics simulations have been performed for a $Cl + Cl_2 \rightarrow Cl_2 + Cl$-like reaction with a 20 kcal/mol^{-1} barrier weakly coupled to a rare gas solvent, [68,69]. Examination of the trajectories shows that those that cross the barrier to reaction have acquired most of the energy necessary to do so via an impulse delivered primarily along the $Cl - Cl_2$ relative motion in the last 200 fs before reaching the barrier top. At 600 fs before reaching the barrier top and at earlier times, the energy distribution and the correlation functions are essentially those of thermalized reagents and solvent. Reaction requires an impulse delivered at the right time and in the right direction. This picture is different from an RRKM-like view of rapid energy exchange among many degrees of freedom (comprising the Cl_3 system and its environment). The motion toward the top of the barrier is not diffusive-like with many small kicks by the solvent to steer the reagents along. The relevant time scale (-200 fs) is such that only a single dominant impulse is delivered.

The second example is an S_N–2-like reaction in water solvent. The simulation [70] is for a $Cl^- + MeCl \rightarrow ClMe + Cl^-$ reaction where Me is a 'methyl atom' of mass 15 and the coupling to the polar water-like solvent by the ionic and dipolar reagents is strong. Given the potential surface and the results of preliminary simulations [71], it appears that vibrational energy in the reactant MeCl bond is needed in both the gas and aqueous phases in order to surmount the barrier. This can be understood by examination of the $(Cl\text{-}Me\text{-}Cl)^-$ potential energy surface, in the absence of solvent, as used in the simulation. The entrance valley runs very parallel to the Cl^--(MeCl) vector with practically no stretch of the MeCl bond. This entrance valley has a well and it ends up in a steeply repulsive wall. In the gas phase, reagents which are not vibrationally excited would come in and rebound from the wall. In the liquid some such trajectories can be trapped in the entrance valley. To follow the very sharp bend in the raction path which leads to the barrier, vibrational excitation is essential. (Just imagine rolling a ball [1] on the potential energy surface shown in figure 1 of [70]. $\beta = 45^{o}$).

4. Vibrational Energy Pathways in Acetylene

Highly vibrationally excited acetylene, in its ground electronic state can be prepared [72] in a two photon resonant process via the electronically excited $\tilde{A}$ state as an intermediate. (This is 'Stimulated Emission Pumping' or SEP, [73]. This double resonance technique results in an essentially purely vibrational spectrum and so eliminates rotational congestion). The equilibrium geometry of the intermediate state is trans-bent. The $C \equiv C$ bond is extended by about 15% as compared to the ground state, while the C-H bond length is hardly changed. At high resolution the spectrum is very dense, with a density of lines roughly comparably to the density of vibrational states that can be accessed. At a somewhat lower resolution the spectrum consists of 'clumps'. At *ca.* 26400 cm^{-1} of excess vibrational energy the density of clumps is about $0.6/cm^{-1}$, [22]. The high density of clumps suggests that they correspond to states with high trans-bend (the ν_4 normal mode) excitation. Since the intermediate $\tilde{A}$ state in the SEP scheme is trans-bent, it is possible that the clumps are the 'bright states' that carry oscillator strength. Our theoretical results do not support this interpretation but this conclusion is quite sensitive to our assumed potential energy surface.

The direct experimental observations establish therefore that there is an intermediate time scale in the problem, corresponding to the clump structure. Taking the Fourier transform of the high resolution spectra establishes [22] that this time is *ca.* 10 ps, corresponding to eigenstates interaction over a range of 3 cm^{-1}. One can also take the Fourier transform of the clump spectrum. The results are unexpected in that there is clear evidence for a short time process of *ca.* 0.5 ps. This corresponds to eigenstates interacting over a range of *ca.* 60 cm^{-1}. There is, therefore, a broad structure in the spectrum which can be clearly discerned upon coarse graining of the experimental frequency spectrum [74]. These broad spectral features, spaced about 200 cm^{-1} apart, are then the primary bright states.

4.1 TRAJECTORY COMPUTATIONS FOR HCCH

An empirical yet realistic potential energy surface was constructed [75] for acetylene in its ground electronic state. A harmonic approximation was used to generate the surface for the excited $\tilde{A}$ state. The 'classical' region on the ground surface which is accessed via SEP was defined as all the points on the lower surface where the potential energy difference to the allowed region on the $\tilde{A}$ surface equals the frequency of the transition. We [75] have chosen a transition corresponding to an excess vibrational energy of 26400 cm^{-1}. An ensemble of classical trajectories corresponding to different initial conditions in the classical region were computed. While these trajectories differed in their detailed time evolution, they mostly were similar in their overall behavior. This was not the case for other trajectories, computed at the same total energy but for quite different initial conditions.

Hence, while most trajectories eventually (i.e., for times longer than *ca.* 15 ps) become chaotic, the route to chaos does depend on the manner of the initial excitation. Even for a tetraatomic molecule at a very high level of excitation, the onset of chaos is not a simple process. This does however mean that our results will be somewhat sensitive to the assumption that the downward transition to the ground state is vertical and classically allowed. This caveat is important and must be borne in mind.

4.2 THE STRETCH MODES

The first modes to respond to the excitation are the two high frequency CH stretches. For over 0.5 ps (several tens of vibrational periods) these are the only modes that are coupled. The other three vibrational modes evolve independently of one another and of the CH stretches, [76].

From the very beginning of local mode models, it was proposed that the two CH modes are coupled via a resonant one quantum exchange [25,26]. It follows that the total number of CH stretch quanta will be conserved. The algebraic approach [77] clearly identifies the corresponding quantum number, denoted by P. States of given P lie in a band (a, so called, multiplet) which is well separated in energy from other multiplets of lower or higher values of P.

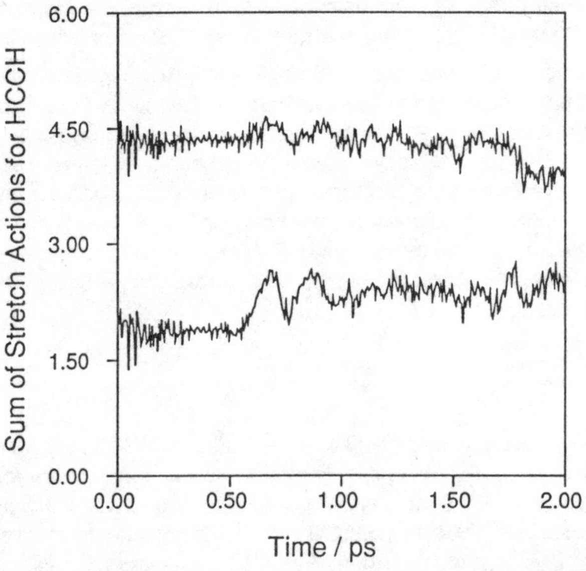

Figure 1. The sum of the classical actions in the two CH modes and in all three stretch modes vs. time. The classical actions are the analogues of the vibrational quantum numbers. Note that in the trajectory computations, the classical P is conserved for over 0.5 ps. Adapted from [78].

The C ≡ C bond is somewhat off-resonance with respect to the C-H stretches. Furthermore, due to the considerably long elongation in the $\widetilde{A}$ state, it is initially highly excited and hence down shifted in frequency. Eventually, it does however couple to the CH stretches via a 1:1 resonance exchange. Then, it is only the sum of the quanta in all three stretch modes, figure 1, which will be conserved.

For DCCD, the C-D stretch frequency is significantly lower hence the coupling to the C ≡ C stretch will be that much more efficient. Indeed, the classical simulations show [78] that while the sum of the C-D stretch actions by itself is not conserved, the sum for all three stretch modes, is. Figure 2 shows the results.

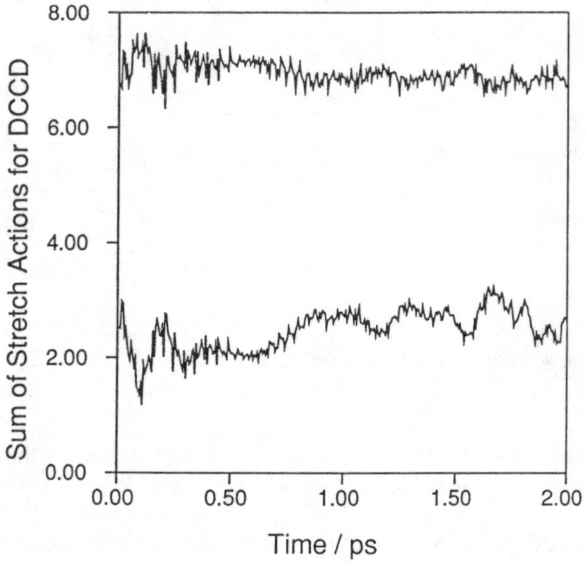

Figure 2. As in figure 1, but for DCCD the lower trace is the classical value of *P*.

Not so evident in figures 1 and 2 is the impulsive nature of the coupling. Hence figure 3 shows the energy in the C ≡ C bond vs. time. Particularly violent are the transfers to the bend past 2.5 ps.

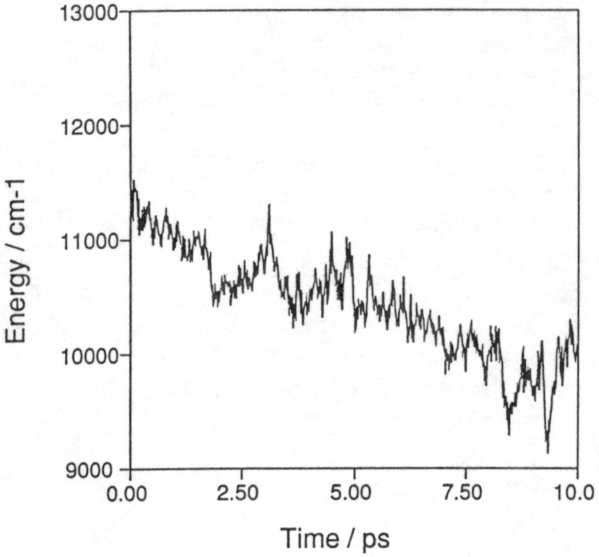

Figure 3. The vibrational energy in the C≡C stretch of HCCH vs. time. Note the high initial value.

It should be reiterated that the conclusion that the CH stretch modes are the first to respond, very much depends on most of the initial conditions in our ensemble corresponding to finite initial excitation, say $P \sim 2$, of these modes. For trajectories corresponding to $P \sim 0$, the CH's remain decoupled from the C≡C mode for much longer and the earliest coupling is that of the C≡C stretch to the bends. Such initial conditions result in a longer time span prior to energy dissipation. A dependence of the dynamics on the initial region sampled in phase space has also been seen in other classical trajectory simulations [79,80].

4.3 THE TRANS-BEND AND THE CLUMPS

For both HCCH and DCCD, the trans-bend excitation (the ν_4 mode, excited because of the geometry of the $\tilde{A}$ surface), does not couple with the other modes for over 3 ps. The sharp spikes in figure 3 beyond 3 ps are due to the onset of this coupling.

Ultimately, the bend is very highly excited and the 'terminal' state for most of the trajectories is for the two hydrogens to 'rotate' in a fairly correlated fashion about the C≡C core. What has not been observed is any significant 'trapping' in a vinyledene-like configuration. (Note however that the energy is about 10 kcalmol^{-1} over the barrier height for the isomerization [81,82].

The coupling of the trans-bend to the stretches leads to a corresponding quantum number in the algebraic approach. We believe that this (approximately conserved) quantum number is the leading to the clump structure in the observed spectra. We also note that to the same level of approximation l_4 is also conserved.

4.4 VINYLEDENE AND THE ORBITING MOTION

Further dynamical details can be obtained by computing time correlation functions, [75]. Taking the Fourier transform one can generate the corresponding frequency spectra. Moreover, one can compute the correlation function only for a time-slice of the trajectory. In this fashion one can discern when certain frequencies contribute to the motion. There are two particularly interesting results that have been obtained in this way. The first is that past the initial coupling of the bend to the C$\equiv$C stretch (which is the energy reservoir, cf. figure 3), the computed frequency spectrum shows a weak peak at a C=C frequency. This is the signature of vinyledene. The area under the peak is however less than 10% of the area of the C$\equiv$C peak. The second result is the nature of the broad frequency response extending from zero to about 10 cm^{-1} with a flat maximum at *ca.* 6 cm^{-1}. This peak is generated by the terminal stage of the motion and corresponds to the rotation of the two hydrogens about the C$\equiv$C core.

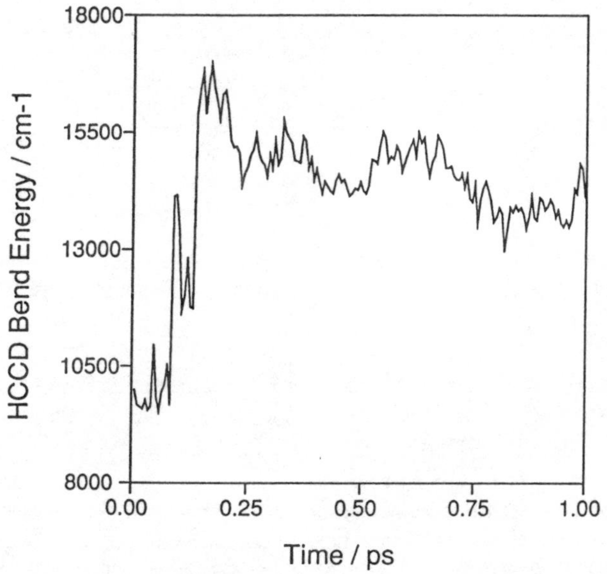

Figure 4. The HCCD bend excitation vs. time. Note the high initial value and its rapid increase.

4.5 HCCD

The computed dynamics for HCCD is quite different from the two symmetric isotopic variants, [78]. From very early on, the bends are strongly coupled to the stretches, figure 4.

The rapid increase in the energy of the HCCD bend is accompanied by a corresponding decline in the C ≡ C stretch excitation, figure 5.

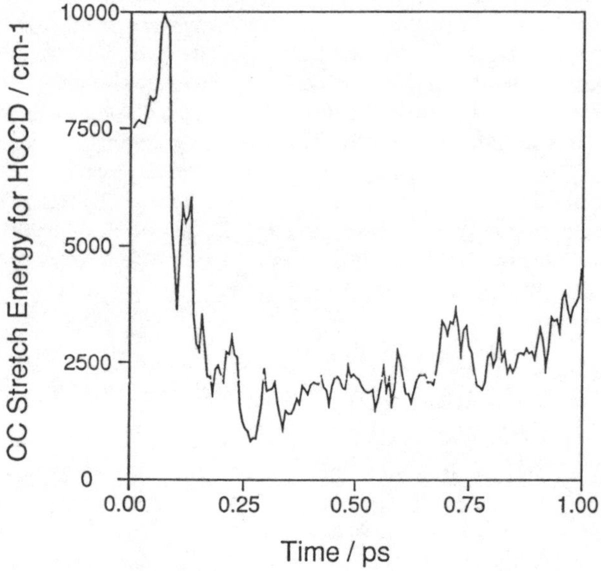

Figure 5. The C ≡ C stretch energy in HCCD vs. time. Ensemble average for the same set of trajectories used to generate figure 4. Note that the gain in energy shown in figure 4 is about compensated by the loss shown here.

The higher frequency C-X, X = H,D modes are about unperturbed during the initial rapid energy exchange between the bend and the C ≡ C stretch. It is only past *ca.* 2 ps that they are coupled to the other modes in a significant fashion.

The dynamical picture for HCCD has similarities with other systems where the large amplitude, soft, motions are the first to respond while the high frequency modes behave adiabatically. Since however all three molecules were computed using the same potential function it is not unreasonable to suggest that the broad and the intermediate spectral features for HCCD may be quite different than those of HCCH and DCCD. In particular, we note that it is possible for HCCH to show a behavior similar to HCCD if the initial conditions (at the same total energy) are drastically altered. It would therefore also be of interest to pump HCCH (or DCCD) via other intermediate states. In particular, the lower is the CH initial excitation (and the higher is the

C ≡ C initial excitation), the more uncoupled are the CH stretches from the other degrees of freedom.

4.6 SUMMARY

The classical trajectory results show a hierarchy of sequential relaxation processes with fairly abrupt transitions from one regime to the next. The purely classical description is consistent with the set of approximate constants of the motion provided by the quantal algebraic approach. The separation of time scales is reflected in the frequency spectrum by features of distinct width, with the short time dynamics corresponding to the broadest features. Not yet subject to experimental test is the possibility that changes in the initial conditions can markedly affect the course of the relaxation nor the predicted isotope effects.

5. Concluding Remarks

Both experimentally and theoretically we are beginning to probe the very act of physical and chemical transformations. Even for systems consisting of many atoms there is now clear evidence that the temporal evolution following the initial excitation cannot be simply described as a rapid dissipation towards equilibrium. Rather, there is often a detailed pathway, with definite experimental signatures.

Very important for highly excited molecules are the large amplitude motions. It is therefore not sufficient to know the potential energy about some minimum energy path. For the case of acetylene, which was discussed in some detail, the theoretical interpretation of the observations for HCCH and the predictions for the isotopic variants are critically dependent on the potential energy function in the region accessed by the stimulated emission pumping and for large H-atom excursions.

Acknowledgment

The work reported reflects the benefit of discussions with many people, R.W. Field, T.A. Holme, J.L. Kinsey, J.P. Pique, K.R. Wilson and R.N. Zare in particular. The computational studies of acetylene were carried out in collaboration with T.A. Holme. The work was supported by the U.S. Air Force Office of Scientific Research, Grant AFOSR 86-0011 and by the United States - Israel Binational Science Foundation. The Fritz Haber Research Center is supported by the Minerva Gesellschaft für die Forschung, mbH, Munich, BRD.

References

1. R.D. Levine and R.B. Bernstein, *Molecular Reaction Dynamics and Chemical Reactivity* (Oxford University Press, N.Y., 1987).
2. I. Oref and B.S. Rabinovitch, *Acc. Chem. Res.* **12**, 166 (1979).
3. H. Reisler and C. Wittig, *Ann. Rev. Phys. Chem.* **37**, 307 (1986); D.W. Lupo and M. Quack, *Chem. Res.* **87**, 181 (1987).
4. M.S. Child, *Chem. Soc. Rev.* **17**, 31 (1988).
5. P. Rogers, D.C. Montage, J.P. Franck, S.C. Tyler and F.S. Rowland, *Chem. Phys. Lett.* **89**, 9 (1982); S.P. Wrigley and B.S. Rabnovitch, *Chem. Phys. Lett.* **98**, 386 (1984); S.M. Lederman, V. Lopez, G.A. Voth and R.A. Marcus, *Chem. Phys. Lett.* **124**, 93 (1986); K.N. Swamy and W.L. Hase, *J. Chem. Phys.* **82**, 123 (1983); T. Uzer and J.T. Hynes in *Stochasticity and Intramolecular Redistribution of Energy*, eds. R. Lefebvre and S. Mukamel, NATO ASI Series (1986).
6. M.J. Rosker, M. Dantus and A.H. Zewail, *Science* **241**, 1200 (1988).
7. D. Imre, J.L. Kinsey, A. Sinha and J. Krenos, *J. Phys. Chem.* **88**, 3956 (1984).
8. H.J. Foth, J.C. Polanyi and H.H. Telle, *J. Chem. Phys.* **86**, 5027 (1982).
9. B.A. Collings, J.C. Polanyi, M.A. Smith, A. Stollow and A.W. Tarr, *Phys. Rev. Lett.* **59**, 2551 (1987).
10. P.R. Brooks, *Chem. Rev.* **88**, 407 (1988).
11. E.J. Heller, *Acc. Chem. Res.* **14**, 368 (1981).
12. E.J. Heller, *J. Chem. Phys.* **72**, 1337 (1986).
13. R. Kosloff, *J. Phys. Chem.* (1988).
14. E.J. Heller, *J. Chem. Phys.* **62**, 1544 (1975).
15. B. Jackson and H. Metiu, *J. Chem. Phys.* **82**, 5707 (1985).
16. R.B. Gerber and M.A. Ratner, *Adv. Chem. Phys.* **70**, 97 (1988); N. Makri and W.H. Miller, J. Chem. Phys. **87**, 5781 (1987).
17. J.D. Doll, R.D. Coalson and D.L. Freeman, *J. Chem. Phys.* **87**, 1641 (1987); J. Chang and W.H. Miller *ibid*, **1648**, (1987).
18. N.Z. Tishby and R.D. Levine, *Phys. Rev.* A**30**, 1477 (1984).
19. J.P. Brunet, C. Leforestier and R.E. Wyatt, *J. Chem. Phys.* **88**, 3125 (1988).
20. Th. Zimmermann, L.S. Cederbaum and H.-D. Meyer, *J. Phys. Chem.* **91**, 4446 (1987).
21. L. Leviander, M. Lombardi, R. Jost and J.-P. Pique, *Phys. Rev. Lett.* **56**, 2449 (1986).
22. P. Pique, Y. Chen, R.W. Field and J.L. Kinsey, *Phys. Rev. Lett.* **58**, 47 (1987).
23. R.D. Levine, *Adv. Chem. Phys.* **70**, 53 (1988).

24. R.D. Levine and J.L. Kinsey **in** *Spectroscopy of Small Molecuels and Ions* (Plenum, N.Y., 1988).

25. M.S. Child and R.T. Lawton, *Faraday Disc. Chem. Soc.* **71**, 233 (1981); I. Halonen, M.S. Child and S. Carter, *Mol. Phys.* **47**, 1097 (1982).

26. B.R. Henry, M.A. Mohammadi and A.W. Tarr, *J. Chem. Phys.* **77**, 3295 (1982).

27. E.L. Sibert, III, W. Reinhardt and J.T. Hynes, *J. Chem. Phys.* **77**, 3583 (1982).

28. R.S. McKay, J.D. Meiss and I.C. Percival, *Physica D***13**, 55 (1984); D. Bensimon and L.P. Kadanoff, *ibid*, D**13**, 82 (1984).

29. R.T. Skodje and M.J. Davis, *J. Chem. Phys.* **88**, 2429 (1988); M.J. Davis, *ibid*, **85**, 1016 (1985).

30. R.B. Shirts and W.P. Reinhardt, *J. Chem. Phys.* **77**, 5204 (1982).

31. D.F. Blintz, D.L. Thompson and J.W. Brady, *J. Chem. Phys.* **86**, 4411 (1987).

32. E.L. Siebert, III., W.P. Reinhardt and J.T. Hynes, *J. Chem. Phys.* **81**, 1115 (1984).

33. V. Buch, R.B. Gerber and M.A. Ratner, *J. Chem. Phys.* **81**, 3393 (1984).

34. K.L. Blintz, D.L. Thompson and J.W. Brady, *J. Chem. Phys.* **86**, 4411 (1987).

35. D. Lu and W.L. Hase, *Chem. Phys. Lett.* **142**, 187 (1987).

36. R.H. Page, Y.R. Shen and Y.T. Lee, *J. Chem. Phys.* **88**, 5562 (1988).

37. W.H. Miller, *Chem. Rev.* **87**, 19 (1987).

38. W.F. Polik, C.B. Moore and W.H. Miller, *J. Chem. Phys.* **89**, 3584 (1988).

39. A. Carrington and R.A. Kennedy, *J. Chem. Phys.* **81**, 91 (1984).

40. J.M. Gomez-Llorente and E. Pollak, *Chem. Phys.* **120**. 37 (1988).

41. M. Berblinger, J.M. Gomez-Llorente, E. Pollak and Ch. Schlier, *Chem. Phys. Lett.* **146**, 353 (1988).

42. M.S. Child, *J. Phys. Chem.* **90**, 3595 (1986).

43. E.W. Schlag, ed., *Intramolecular Processes, Ber. Bunsenges. Phys. Chem.* (1988).

44. J. Kommandeur, *Adv. Chem. Phys.* **70**, 133 (1988); P.M. Felker and M.A. Zewail, *ibid*, **70**, 265 (1988).

45. J. Jortner and R.D. Levine, *Adv. Chem. Phys.* **47**, 1 (1981).

46. G.E. Ewing, *J. Phys. Chem.* **91**, 4662 (1987).

47. D.H. Levy, *Adv. Chem. Phys.* **47**, 241 (1981).

48. J.A. Beswick and J. Jortner, *Adv. Chem. Phys.* **47**, 363 (1981).

49. S.B. Woodruff and D.L. Thompson, *J. Chem. Phys.* **71**, 376 (1979).

50. R.B. Gerber, V. Buch and M.A. Ratner, *J. Chem. Phys.* **77**, 3022 (1982).

51. M.J. Davis and S.K. Gray, *J. Chem. Phys.* **84**, 5389 (1986); S.K. Gray and S.A. Rice, *Faraday Discuss. Chem. Soc.* **82** (1986).

52. D.A. Dixon, D.R. Herschbach and W. Klemperer, *Faraday Discuss. Chem. Soc.* **62**, 341 (1977).

53. V. Buch, R.B. Gerber and M.A. Ratner, *Chem. Phys. Letters* **101**, 44 (1983); M.A. Ratner and R.B. Gerber, *J. Phys. Chem.* **90**, 20 (1986).

54. J.W. Brady, J.D. Doll and D.L. Thompson, *J. Chem. Phys.* **71**, 2467 (1979); *ibid.* **73**, 2767 (1980).

55. F.G. Amar and B.J. Berne, *J. Phys. Chem.* **88**, 6720 (1984).

56. I. NoorBatcha, L.M. Raff and D.L. Thompson, *J. Chem. Phys.* **81**, 5658 (1984).

57. D.R. Worsnop, S.J. Buelow and D.R. Herschbach, *J. Phys. Chem.* **90**, 5121 (1986).

58. D. Lucas and G.E. Ewing, *Chem. Phys.* **58**, 385 (1983).

59. R.R. Lucchese and J.C. Tully, *J. Chem. Phys.* **81**, 6313 (1984); C. Lim and J.C. Tully, *ibid*, **85**, 7423 (1986); R.R. Lucchese, *J. Chem. Phys.* **86**, 443 (1987).

60. A.C. Beri and T.F. George, *J. Chem. Phys.* **87**, 4147 (1987); T.F. George, K.T. Lee, W.C. Murphy, M. Hutchinson and H.W. Lee **in** *Theory of Chemical Reaction Dynamics*, M. Baer, ed., CRC Press.

61. R.B. Hall, *J. Phys. Chem.* **91**, 1007 (1987).

62. A.A. Deckert and S.M. George, *Surf. Sci.* **182**, L215 (1987).

63. R.N. Zare and R.D. Levine, *Chem. Phys. Letters* **136**, 593 (1987).

64. J.H. Hahn, R. Zenobi and R.N. Zare, *JACS* **109**, 2842 (1987).

65. J. Grotemeyer, U. Bosel, K. Walter and E.W. Schlag, *JACS* **108**, 4233 (1986).

66. T.A. Holme and R.D. Levine, to be published.

67. T.A. Holme and R.D. Levine, *Faraday Discuss. Chem. Soc.* (1987).

68. J.P. Bergsma, J.R. Reimers, K.R. Wilson and J.T. Hynes, *J. Chem. Phys.* **85**, 5625 (1986).

69. I. Benjamin, B.J. Gertner, N.J. Tang and K.R. Wilson, to be published.

70. J.P. Bergsma, B.J. Gertner, K.R. Wilson and J.T. Hynes, *J. Chem. Phys.* **86**, 1356 (1987).

71. I. Benjamin, B.J. Getner, J.H. Kwak and K.R. Wilson, to be published.

72. E. Abramson, R.W. Field, D. Imre, K.K. Ines and J.L. Kinsey, *J. Chem. Phys.* **83**, 453 (1985).

73. C.E. Hamilton, J.L. Kinsey and R.W. Field, *Ann. Rev. Phys. Chem.* **37**, 493 (1986).

74. J.P. Pique, Y.M. Engel, R.D. Levine, Y. Chen, R.W. Field and J.L. Kinsey, *J. Chem. Phys.* **88**, 5972 (1988).

75. T.A. Holme and R.D. Levine, to be published.

76. T.A. Holme and R.D. Levine, *J. Chem. Phys.* **89**, 3379 (1988).

77. (a) O.S. van Roosmalen, I. Benjamin and R.D. Levine, *J. Chem. Phys.* **81**, 5986 (1984); (b) I. Benjamin, O.S. van Roosmalen and R.D. Levine, *ibid* **81**, 3352 (1984).

78. T.A. Holme and R.D. Levine, *Chem. Phys. Lett.* **150**, 393 (1988).

79. S.C. Farantos, *J. Chem. Phys.* **85**, 641 (1986).

80. B.G. Sumpter and D.L. Thompson, *J. Chem. Phys.* **86**, 2805 (1987).

81. S.K. Gray, W.H. Miller, Y. Yamaguchi and H.F. Schaefer III, *J. Am. Chem. Soc.* **103**, 1900 (1981); T. Carrington, Jr., L.M. Hubbard, H.F. Schaefer III and W.H. Miller, *J. Chem. Phys.* **80**, 4347 (1984).

82. J.A. Pople, K. Raghavachari, M.J. Frisch, J.S. Binkley and P.v.R. Schleyer, *J. Am. Chem. Soc.* **105**, 6389 (1983).

QUANTUM MECHANICS OF CHEMICAL REACTIONS: RECENT DEVELOPMENTS IN REACTIVE SCATTERING AND IN REACTION PATH HAMILTONIANS

William H. Miller
Department of Chemistry, University of California, and Materials and Chemical Sciences Division, Lawrence Berkeley Laboratory, Berkeley, California 94720 USA

ABSTRACT. Two recent developments in the theory of chemical reaction dynamics are reviewed. First, it has recently been discovered that the S-matrix version of the Kohn variational principle is free of the "Kohn anomalies" that have plagued other versions and prevented its general use. This has considerably simplified quantum mechanical reactive scattering calculations, which provide the rigorous characterizations of bimolecular reactions. Second, a new kind of reaction path Hamiltonian has been developed, one based on the "least motion" path that interpolates linearly between the reactant and product geometry of the molecule (rather than the previously used minimum energy, or "intrinsic" reaction path). The form of Hamiltonian which results is much simpler than the original reaction path Hamiltonian, but more important is the fact that it provides a more physically correct description of hydrogen atom transfer reactions.

1. INTRODUCTION

In this paper I will review some of the recent developments in the theory of chemical reactions. The first topic, quantum reactive scattering, pertains to the most rigorous theoretical description (i.e., explicit solution of the Schrödinger equation with appropriate boundary conditions) of a chemical reaction. Not surprisingly, this methodology is currently applicable only to the simplest chemical processes (but in full 3-dimensional space), e.g., an atom-diatom reaction, A+BC → AB+C. It is nevertheless exciting to see that it is now becoming possible to carry out the rigorous quantum calculations (i.e., a "simulation") that characterize these simplest reactions to the most complete level of detail allowed by the laws of nature.

The second topic deals with the theoretical description of more complex chemical systems. Specifically, a new class of reaction path models is described, namely the <u>diabatic</u> reaction path Hamiltonian. The difference between this new model and the earlier reaction path Hamiltonian is that the present one is based on a <u>least motion</u>

J. Jortner and B. Pullman (eds.), Perspectives in Quantum Chemistry, 57–82.

reference path (i.e., a path that is linear interpolation between reactants and products) rather than on the minimum energy path the ("intrinsic" reaction path). This new version is more useful for treating hydrogen atom transfer reactions than is the earlier one (which is useful for many other kinds of reactions).

Before beginning the discussion of reactive scattering in Section 2, it is perhaps useful to take a few paragraphs here in the Introduction to summarize the background to these developments. In 1969 a general formulation of quantum scattering for chemical reactions was presented[1] which was a natural generalization of earlier work in electron scattering. The novel feature was that the wavefunction is expressed as a coupled channel expansion in standard Jacobi coordinates, but in all arrangements (i.e., A+BC, AB+C, AC+B). Coupling between states of different arrangements leads to a non-local, i.e., exchange-type of interaction, and this is what makes reactive scattering difficult in this formulation. (This reactive exchange interaction is analogous to electron exchange interactions that result when the electronic wavefunctions is antisymmetrized, i.e., expressed as linear combinations of different "arrangements" of the electrons.) Some other formulations[2] of chemically reactive scattering avoid these exchange interactions, which is of course an advantage, but they have other kinds of disadvantages of their own.

The only general way to deal with these exchange interactions seems to be[3] to expand the dependence of the wavefunction on the scattering coordinates in a basis set, using a variational principle to determine the expansion coefficients. Several such variational principles exist,[4] and they all work, but the simplest one to apply is the Kohn variational principle;[5] this is because it involves matrix elements only of the Hamiltonian operator itself and not those involving the Green's function for a reference problem. The Kohn principle has not been of general use, however, because of "Kohn anomalies",[6,7] i.e., spurious singularities that appear in the energy dependence of the scattering results.

The important recent discovery, however, is that there are no such "anomalies" if the Kohn principle is applied with S-matrix-type boundary conditions[8] (as opposed to K-matrix boundary conditions). With this rather subtle feature of the Kohn principle now understood, it provides a reliable and extremely straight-forward approach to quantum scattering, equally applicable to reactive or non-reactive processes. Section 2 describes this S-matrix version of the Kohn variational principle as it applies to chemically reactive scattering. This S-matrix version of the Kohn method has also been recently applied to electron-atom/molecule scattering,[9] with excellent results.

Mention should also be made of a number of recent reactive scattering calculations by Kouri, Truhlar, and coworkers.[10] These workers employ the coupled operator formalism of Baer and Kouri[11] (though the version of it they use makes it identical to the formulation of ref. 1), and then use the Newton variational method[12] for the amplitude density[13] to solve the equations. These authors have obtained excellent results for reaction probabilities of several atom-diatom reactions ($H+H_2$, $O+H_2$, $Br+H_2$) for zero total angular momentum (J=0), though we note the obvious disadvantage that this approach requires matrix elements of the operator ($G_o-G_oVG_o$).

The new diabatic reaction path Hamiltonian[14] is described in Section 3, and more background and the motivation for its development is given there. Its most notable feature is that there are no "curvature couplings" as in the original reaction path Hamiltonian[15] - because the reference path is straight - and the coriolis couplings between different modes are also eliminated. The kinetic energy is thus completely Cartesian-like. The price for this simplification of the kinetic energy is that the quadratic part of the potential energy now has off-diagonal terms. (There is also a term in the potential energy that is linear in the "bath" modes because the reference path is not the minimum energy path.) This elimination of kinetic energy coupling, at the expense of introducing coupling into the potential energy, is analogous to the diabatic electronic representation of a vibronic Hamiltonian and is the reason for our use of the term. As in the vibronic case, it is often easier to deal with the dynamics when the coupling appear in the potential rather than the kinetic energy.

2. S-MATRIX VERSION OF THE KOHN VARIATIONAL PRINCIPLE

2.1. General Methodology

All relevant features of the methodology are illustrated by simple s-wave potential scattering. It[8b] will thus first be described with regard to this problem, and the generalization to multichannel rearrangement scattering given at the end.

The Hamiltonian is of the standard form

$$H = \frac{-\hbar^2}{2\mu}\frac{d^2}{dr^2} + V(r), \tag{2.1}$$

where $V(r) \to 0$ as $r \to \infty$. The S-matrix version of the Kohn variational approximation to the S-matrix (at energy E) can be stated as

$$S = \text{ext}[\tilde{S} + \frac{i}{\hbar}\langle\tilde{\psi}|H-E|\tilde{\psi}\rangle], \tag{2.2}$$

where $\tilde{\psi}(r)$ is a trial wavefunction that is regular at r=0 and has asymptotic form (as $r \to \infty$)

$$\tilde{\psi}(r) \sim -e^{-ikr}v^{-1/2} + e^{ikr}v^{-1/2}\tilde{S} \tag{2.3}$$

where $v=\hbar k/\mu$ is the asymptotic velocity. (Note: The convention is used throughout this paper that the wavefunctions in the bra symbol

< | in bra-ket matrix element notation are not complex conjugated.) "ext" in Eq. (2.2) means that the quantity in square brackets is to be extremized by varying any parameters in $\psi(r)$. (Note that for a given trial function ψ, Eq. (2.2) may also be viewed as the distorted wave Born approximation, where ψ is the distorted wave.)

A linear variational form is taken for the trial function $\tilde{\psi}(r)$,

$$\tilde{\psi}(r) = -u_0(r) + \sum_{\ell=1}^{N} u_\ell(r)c_\ell, \tag{2.4}$$

where $u_0(r)$ is a function that is regular at r=0 and has the asymptotic form (as $r \to \infty$)

$$u_0(r) \sim e^{-ikr} v^{-\frac{1}{2}}. \tag{2.5}$$

A simple choice for $u_0(r)$ is

$$u_0(r) = f(r)e^{-ikr} v^{-\frac{1}{2}}, \tag{2.6}$$

where f(r) is a smooth cut-off function,

$$\begin{aligned} f(r) &\to 0 \ , \ r \to 0 \\ f(r) &\to 1 \ , \ r \to \infty, \end{aligned} \tag{2.7}$$

such as $f(r)=1-e^{-\alpha r}$. (More generally, $u_0(r)$ may be the (irregular) solution of some (e.g., long range) distortion potential that has asymptotic form Eq. (2.5), multiplied by a cut-off function to regularize it at r=0). The function $u_1(r)$ is

$$u_1(r) = u_0(r)^*, \tag{2.8}$$

and the basis functions $\{u_\ell(r)\}$, $\ell=2,\ldots,N$ are real, square-integrable functions. The coefficients $\{c_\ell\}$, $\ell=1,\ldots,N$ in Eq. (2.4) are the variational parameters in ψ.

With ψ of Eq. (2.4) substituted into Eq. (2.2) and the coefficients $\{c_\ell\}$ varied to extremize it, one obtains the following expression for the S-matrix

$$S = \frac{i}{\hbar} (M_{0,0} - \underset{\sim}{M}_0^T \cdot \underset{\approx}{M}^{-1} \cdot \underset{\sim}{M}_0), \tag{2.9}$$

where $M_{0,0}$ is a 1x1 "matrix", $\underset{\sim}{M}_0$ a 1xN matrix, and $\underset{\approx}{M}$ an NxN matrix,

$$M_{0,0} = \langle u_0|H-E|u_0\rangle \qquad (2.10a)$$

$$(\underset{\sim}{M}_0)_\ell = \langle u_\ell|H-E|u_0\rangle \qquad (2.10b)$$

$$(\underset{\approx}{M})_{\ell,\ell'} = \langle u_\ell|H-E|u_{\ell'}\rangle\ , \qquad (2.10c)$$

for $\ell,\ell'=1,\ldots,N$, and where "T" denotes matrix transpose. Note that all matrix elements involving the unbounded basis functions u_0 and u_1 exist because

$$\lim_{r\to\infty} (H-E) \begin{cases} u_0(r) \\ u_1(r) \end{cases} = 0. \qquad (2.11)$$

Since the matrix $\underset{\approx}{M}$ of Eq. (2.10c) is complex-symmetric, there are no real values of E for which the matrix inverse in Eq. (2.9) is singular, and thus no "Kohn anomalies". In fact, the condition that Eq. (2.9) is singular, namely

$$\det(\underset{\approx}{M}) = \det\,[\langle u_\ell|H-E|u_{\ell'}\rangle] = 0, \qquad (2.12a)$$

$\ell,\ell'=1,\ldots,N$, is the secular equation for eigenvalues of the Schrödinger equation

$$(H-E)\psi(r) = 0, \qquad (2.12b)$$

with boundary condition (as $r\to\infty$)

$$\psi(r) \propto e^{ikr}. \qquad (2.12c)$$

I.e., Eq. (2.12a) is the expression that has been used before[16] for determining Siegert eigenvalues,[17] the complex energies that are the (physically correct) complex poles of the S-matrix which characterize the positions and widths of scattering resonances. Eq. (2.9) is thus singular only where it is supposed to be singular.

For comparison, the Kohn principle for the K-matrix gives a similar expression,[7]

$$K = -\frac{2}{\hbar}\left(\tilde{M}_{0,0}-\underset{\sim}{\tilde{M}}_0{}^{T}\cdot\underset{\approx}{\tilde{M}}{}^{-1}\cdot\underset{\sim}{\tilde{M}}_0\right), \qquad (2.13a)$$

where the matrix elements here have the same form as Eq. (2.10) except that the function $\tilde{u}_0$ and $\tilde{u}_1$ are different,

$$\tilde{u}_0(r) = -\mathrm{Im}\ u_0(r) \sim \frac{\sin kr}{v^{1/2}} \qquad (2.13b)$$

$$\tilde{u}_1(r) = \mathrm{Re}\ u_0(r) \sim \frac{\cos kr}{v^{1/2}}. \qquad (2.13c)$$

The fact the matrix $\tilde{\underset{\approx}{M}}$ is real and symmetric leads to real values of E for which

$$\det(\tilde{\underset{\approx}{M}}) = 0, \qquad (2.14)$$

and thus real values of E for which Eq. (2.13a) is singular, i.e., the Kohn anomalies.[18]

To emphasize again, use of the Kohn variational principle with standing wave boundary conditions to obtain (an approximation to) the K-matrix, as in Eq. (2.13), and then the S-matrix via the relation $S = (1+iK)(1-iK)^{-1}$, is not equivalent to using the Kohn principle with running wave boundary conditions to obtain (an approximation to) the S-matrix directly, i.e., Eqs. (2.2)-(2.10). And furthermore, as discussed above, the latter procedure is free of anomalous singularities and thus the preferred version of the Kohn method.

The S-matrix Kohn approach also allows one to identify a corresponding basis set approximation to matrix elements of the full outgoing wave Green's function $G^+(E) \equiv (E+i\varepsilon-H)^{-1}$. This is[8a]

$$\langle a|G^+(E)|b\rangle = -\sum_{\ell,\ell'=1}^{N} \langle a|u_\ell\rangle\, (\underset{\approx}{M}^{-1})_{\ell,\ell'} \langle u_{\ell'}|b\rangle, \qquad (2.15)$$

where $\underset{\approx}{M}$ is as above, Eq. (2.10c), and $|a\rangle$ and $|b\rangle$ are any square-integrable functions. Note that the complex-symmetric structure of the matrix $\underset{\approx}{M}$ is the same as that in complex scaling/coordinate rotation theory,[19-22] and for the same reasons. If the functions $|a\rangle$ and $|b\rangle$ are real, then Eq. (2.15) leads to a useful way for calculating matrix elements of the density of states operator,

$$\langle a|\delta(E-H)|b\rangle = -\pi^{-1}\mathrm{Im}\langle a|G^+(E)|b\rangle. \qquad (2.16)$$

In actual calculations for the S-matrix, Eq. (2.9), one does not wish to carry out numerical calculations with the complex symmetric matrix $\underset{\approx}{M}$. This can be avoided by the usual partitioning methods, so that Eq. (2.9) can be written in the equivalent form

$$S = \frac{i}{\hbar}\,(B - C\cdot B^{*-1}\cdot C), \qquad (2.17)$$

$$u_{0n}^{\gamma}(r_\gamma) \sim e^{-ik_{n\gamma}r_\gamma}/v_{n\gamma}^{1/2} .$$

$\underset{\approx}{M}$ is a "large" by "large" real symmetric matrix in the composite space, internal plus translational,

$$(\underset{\approx}{M})_{\ell n\gamma,\ell'n'\gamma'} = \langle u_{\ell n}^{\gamma}\phi_n^{\gamma}|H-E|u_{\ell'n'}^{\gamma'}\phi_{n'}^{\gamma'}\rangle, \qquad (2.20c)$$

where $\{u_{\ell n}^{\gamma}(r_\gamma)\}$ is a square integrable basis (that need not depend on n - i.e., the same translational basis can be used for every channel). $\underset{\approx}{M}_0$ is a "large" by "small" rectangular matrix

$$(\underset{\approx}{M}_0)_{\ell n\gamma,n'\gamma'} = \langle u_{\ell n}^{\gamma}\phi_n^{\gamma}|H-E|u_{0n'}^{\gamma'}\phi_{n'}^{\gamma'}\rangle. \qquad (2.20d)$$

Only open channels $\{n\gamma\}$ are included in the matrices $\underset{\approx}{M}_{00}$, $\underset{\approx}{M}_{1,0}$, and the "small" dimension of $\underset{\approx}{M}_0$, while open and closed channels are required in the matrix $\underset{\approx}{M}$ and the "large" dimension of $\underset{\approx}{M}_0$.

Eqs. (2.19) - (2.20) thus express the S-matrix for reactive scattering in an extremely straight-forward manner: one chooses basis functions, computes matrix elements of the Hamiltonian, and then does a standard linear algebra calculation.

2.2. Application to $F+H_2$

Initial application[8c] of the above methodology was made to the standard benchmark problem, the 3-d $H+H_2 \rightarrow H_2+H$ reaction for J (the total angular momentum quantum number) = 0. The results showed the S-matrix version of the Kohn method to be accurate, efficient, and stable.

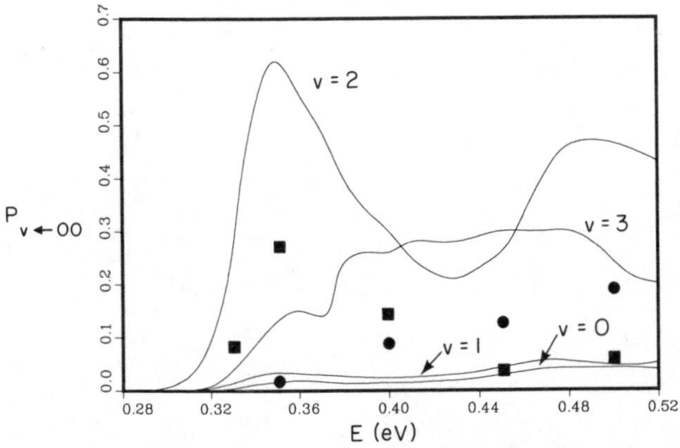

Figure 1. Reaction probabilities for $H_2(v=j=0)+F \rightarrow HF(v)+H$, summed

where B and C are the 1x1 "matrices"

$$B = M_{0,0} - \underset{\sim}{M}_0^T \cdot \underset{\approx}{M}^{-1} \cdot \underset{\sim}{M}_0 \tag{2.18a}$$

$$C = M_{1,0} - \underset{\sim}{M}_0^{*T} \cdot \underset{\approx}{M}^{-1} \cdot \underset{\sim}{M}_0, \tag{2.18b}$$

when $M_{0,0}$, $\underset{\sim}{M}_0$, and $\underset{\approx}{M}$ are as before, Eq. (2.10), except that $\ell,\ell' = 2, \ldots, N$ (i.e., only the real basis functions), and

$$M_{1,0} = \langle u_0^* | H-E | u_0 \rangle. \tag{2.18c}$$

Here the matrix $(\underset{\approx}{M})_{\ell,\ell'}$, $\ell,\ell' = 2, \ldots, N$ is real and symmetric, and thus more easily dealt with. (One can readily verify that a value of E for which $\det(\underset{\approx}{M}) = 0$ does not lead to a singularity in Eq. (2.17.)

Finally, for general multichannel rearrangement scattering,[1] let $(\underset{\sim}{q}_\gamma, r_\gamma)$ denote the internal coordinates and radial scattering (i.e., translational) coordinate for arrangement γ; $\{\phi_n^\gamma(\underset{\sim}{q}_\gamma)\}$ are the asymptotic channel eigenfunctions for the internal degrees of freedom. Eqs. (2.17) and (2.18) generalize as follows

$$\underset{\approx}{S} = \frac{i}{\hbar} (\underset{\approx}{B} - \underset{\approx}{C}^T \cdot \underset{\approx}{B}^{*-1} \cdot \underset{\approx}{C}), \tag{2.19a}$$

where $\underset{\approx}{S}$, $\underset{\approx}{B}$, and $\underset{\approx}{C}$ are "small" square matrices, the dimension of the number of open channels, e.g., $\underset{\approx}{S} = [S_{n\gamma,n'\gamma'}]$, etc. $\underset{\approx}{B}$ and $\underset{\approx}{C}$ are given by

$$\underset{\approx}{B} = \underset{\approx}{M}_{0,0} - \underset{\approx}{M}_0^T \cdot \underset{\approx}{M}^{-1} \cdot \underset{\approx}{M}_0 \tag{2.19b}$$

$$\underset{\approx}{C} = \underset{\approx}{M}_{1,0} - \underset{\approx}{M}_0^{*T} \cdot \underset{\approx}{M}^{-1} \cdot \underset{\approx}{M}_0, \tag{2.19c}$$

where $\underset{\approx}{M}_{0,0}$ and $\underset{\approx}{M}_{1,0}$ are also "small" square matrices

$$(\underset{\approx}{M}_{0,0})_{n\gamma,n'\gamma'} = \langle u_{0n}^\gamma \phi_n^\gamma | H-E | u_{0n'}^{\gamma'} \phi_{n'}^{\gamma'} \rangle \tag{2.20a}$$

$$(\underset{\approx}{M}_{1,0})_{n\gamma,n'\gamma'} = \langle u_{0n}^{\gamma *} \phi_n | H-E | u_{0n'}^{\gamma'} \phi_{n'}^{\gamma'} \rangle, \tag{2.20b}$$

$u_{0n}^\gamma(r_\gamma)$ is a function regular at $r_\gamma = 0$ and with asymptotic form (as $r_\gamma \to \infty$), over final rotational states of HF, for total angular momentum J=0, as a function of total energy (relative to the minimum of the potential energy surface in the reactant valley). The squares (v=2) and circles (v=3) are results of earlier, less accurate calculations of other workers.

Much more impressive, though, is the calculation[8d] for the $F+H_2 \rightarrow HF+H$ reaction, also for J=0. Because the reaction is 32 kcal/mole exothermic, there are many HF vibrational and rotational states that must be included in the coupled channel expansion. Fig. 1 shows the reaction probabilities (the square of the S-matrix elements) for ground state (v=j=0) H_2 and various final vibrational states of HF (summed over final rotational states). These are the first quantitative calculations for the 3-d version of this reaction, and one sees the well-known population inversion that leads to the HF chemical laser.

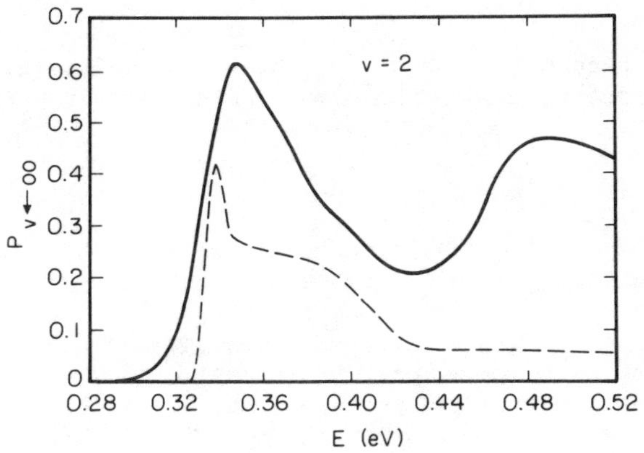

Figure 2. The HF(v=2) result of Fig. 1 (solid curve) compared to the collinear reaction probability (dashed curve) $H_2(v=0)+F \rightarrow HF(v=2)+H$ of ref. 23, using the same potential energy surface. The collinear result has been shifted in energy by $\hbar\omega_b^{\ddagger}$, $\omega_b^{\ddagger}$ being the bending frequency of H-H-F at the saddle point on the potential energy surface.

Fig. 2 shows the 0→2 reaction probability compared to the analogous collinear result,[23] the latter having been shifted in energy by $\hbar\omega_b^{\ddagger}$, $\omega_b^{\ddagger}$ being the bending frequency of the "activated complex". For reactions that have strongly collinear potential energy surfaces (i.e., large $\hbar\omega_b^{\ddagger}$) one often finds[24] that the energy-shifted collinear reaction probability is a good approximation to the corresponding 3-d vibrational reaction probabilities (summed over rotational states). This is seen not to be the case here, presumably because the bending frequency at the transition state is very small for $F-H_2$. Finally, Fig. 3 shows the distribution of final rotational states that results from the $F+H_2(v=0) \rightarrow HF(v=2)+H$ transition. One sees considerable rotational excitation in the products.

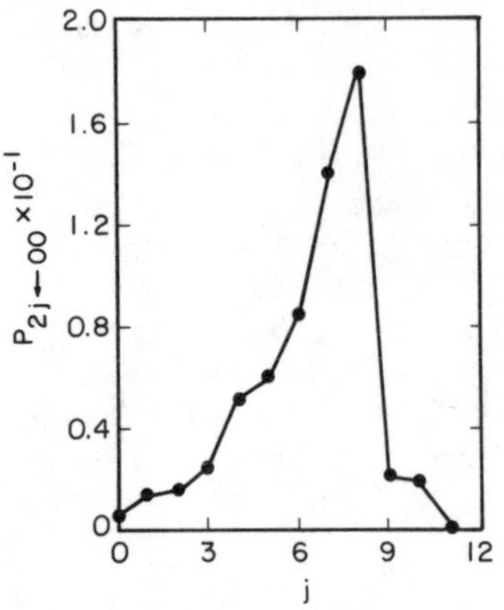

Figure 3. Reaction probability for $H_2(v=j=0)+F \rightarrow HF(v=2,j)+H$, as a function of j, the rotational quantum number of HF, also for total angular momentum J=0.

2.3. Enhancements

There are a number of further developments and enhancements that should make the S-matrix version of the Kohn variational principle even more useful for reactive (and electron-molecule) scattering. One of these is use of a discrete variable,[25] or pseudo-spectral representation[26] for the translational basis set. Though this might require the use of slightly larger translational bases, it has the very desirable feature that matrix elements of the potential energy are diagonal, i.e., simply the value of the potential function at the grid points. Thus rather than having to evaluate matrix elements by numerical integration for a 5000x5000 potential energy matrix, say, one has only to evaluate the potential energy function at 5000 points.

The idea of basis set contraction that is commonly used in computational quantum chemistry[27] should also prove useful for the present calculation. Thus to reduce the size of the transitional basis, one diagonalizes a zeroth matrix in the translational index alone and then chooses a sub-set of these eigenfunctions ("better" basis functions) as the translational basis for the full calculation.

Another extremely important simplification has been pointed out by Rescigno and Schneider,[28] namely that it is essentially no approximation to neglect exchange (i.e., $\gamma\neq\gamma'$) matrix elements in the "free-free" matrices $\underset{\approx}{M}_{00}$ and $\underset{\approx}{M}_{1,0}$, Eqs. (2.20a and b), and in the "bound-free" matrix $\underset{\sim}{M}_0$, Eq. (2.20d). It is clear that this will be possible for the present application[29] because the "free" translational functions $u_{0n}^{\gamma}(r_\gamma)$ include a cut-off function that cause them to vanish in the close-in interaction region where exchange is significant. Thus the exchange interaction is mediated entirely by the "bound-bound" matrix $\underset{\approx}{M}$, Eq. (2.20c). The practical significance of this is that "direct" matrix elements (the ones diagonal in the arrangement index γ) are much easier to compute than exchange ones, and furthermore it is the matrices $\underset{\approx}{M}_{00}$, $\underset{\approx}{M}_{1,0}$, and $\underset{\approx}{M}_0$ that must be recalculated anew at each scattering energy (because $u_{0n}^{\gamma}(r_\gamma)$ is energy-dependent). This observation thus considerably simplifies the energy-dependent part of the calculation.

Another strategy that may be useful is that suggested parenthetically after Eq. (2.7), i.e., to incorporate distorted wave-like information into the function $u_0(r)$ (and $u_1 \equiv u_0^*$). The most complete version of this idea would be to use a multichannel distorted wave[30] for u_0. More specifically, consider Eqs. (2.19)-(2.20) for the general multichannel rearrangement case. One modifies the function $u_{0n}(r)$ in the following way

$$u_{0n}(r)\phi_n(\underset{\sim}{q}) \rightarrow \sum_{n''} \phi_{n''}(\underset{\sim}{q}) u^{(0)}_{n''\leftarrow n}(r) f(r), \tag{2.21}$$

where f(r) is a cut-off function as before, and $u^{(0)}_{n''\leftarrow n}(r)$ satisfies the open channel inelastic coupled-channel Schrödinger equation with asymptotic boundary condition

$$u^{(0)}_{n''\leftarrow n}(r) \sim \delta_{n'',n}\, e^{-ik_n r}\, v_n^{-1/2}. \tag{2.22}$$

(The arrangement index γ has been dropped here because, as noted in the above paragraph, we do not need to consider matrix elements with these functions between different arrangements.) In practice one determines the functions $\{u^{(0)}_{n''\leftarrow n}(r)\}$ by beginning at large r with the initial condition of Eq. (2.22) and integrating the inelastic (non-reactive) coupled channel equations inward as far as is needed; the cut-off function f(r) determines how far in this is. The matrices $\underset{\approx}{M}_{0,0}$ and $\underset{\approx}{M}_{1,0}$ of Eqs. (2.20a) and (2.20b) (now diagonal in γ) can be shown then to take the very simple form

$$\left(\underset{\approx}{M}_{0,0}\right)_{n,n'} = \frac{\hbar^2}{2\mu} \sum_{n''} \langle u^{(0)}_{n''\leftarrow n} | f'^2 | u^{(0)}_{n''\leftarrow n'} \rangle \tag{2.23a}$$

$$\left(\underset{\approx}{M}_{1,0}\right)_{n,n'} = \frac{i\hbar}{2}\,\delta_{n,n'} + \sum_{n''} \langle u^{(0)*}_{n''\leftarrow n} | f'^2 | u^{(0)}_{n''\leftarrow n'} \rangle \tag{2.23b}$$

for each arrangement γ. The rectangular matrix takes a correspondingly simple form. The virtue of using this more sophisticated function for u_0 is that the short-range basis functions $\{u^{\gamma}_{\ell n}\}$ now need to span a much smaller region of space so that fewer of them will be required. Applications using this approach are in progress.[31]

Finally, since it is the S-matrix that is being calculated, one has the option of computing the full matrix or only one row of it. If a general purpose (e.g., LU decomposition) algorithm is used to evaluate $\underset{\approx}{M}^{-1}\cdot\tilde{M}_0$ in Eq. (2.19), then there is little economy in evaluating only one row of the S-matrix. If iterative methods (e.g., Lanczos recursion[32]) are used, however, the effort is proportional to the number of rows of the S-matrix that one evaluates. If one is interested in only one, or a few initial states, then such a procedure will be considerably more efficient, meaning that substantially larger calculations will be feasible.

3. A DIABATIC REACTION PATH HAMILTONIAN

3.1. Background

The idea of a reaction path is a venerable one in the theory of chemical reactions.[33,34] The minimum energy reaction path on the Born-Oppenheimer potential energy surface, also called the intrinsic reaction path,[34d] is uniquely defined as the steepest descent path (in mass-weighted Cartesian coordinates) from the transition state (the saddle point on the potential surface) down to the local minima that are the equilibrium geometries of reactants and products. More recently[15] it was shown how to express the (classical or quantum) Hamiltonian of an N atom molecular system in terms of the reaction coordinate, the distance along this reaction path, and 3N-7 local normal mode coordinates for vibrations orthogonal to it (and three Euler angles for overall rotation of the N atom system), plus momentum variables (or operators) conjugate to these coordinates. This reaction path Hamiltonian has been used successfully to describe a variety of processes in polyatomic reaction dynamics.[35,36]

Though the reaction path Hamiltonian based on the minimum energy path has proved useful for many reactions, and will surely do so for many others, there are situations for which it is not appropriate. One of the most important of these is H-atom transfer reactions, a prototype of which is the symmetric H-atom transfer in malonaldehyde,[37]

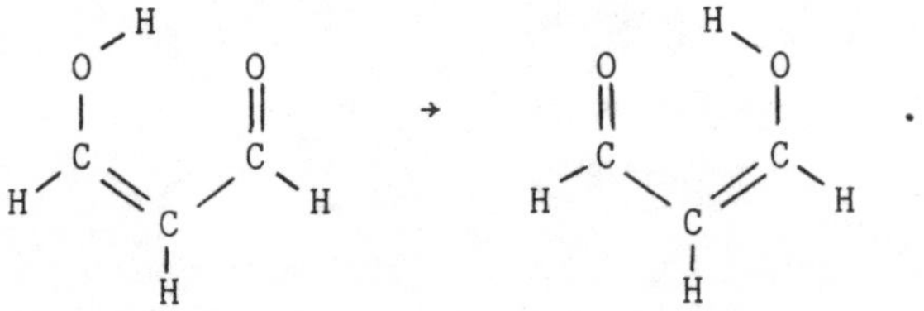

(3.1)

This is a polyatomic version of a heavy + light-heavy mass combination reaction, for which the proto-type is a simple atom-diatom reaction such as

$$C\ell + HC\ell \rightarrow C\ell H + C\ell. \tag{3.2}$$

For this atom-diatom system it is well-known[38] that the minimum energy path is very sharply curved, so that the relevant dynamical motion deviates far from it. It is also well-known that the reaction path Hamiltonian (which reduces to Marcus' natural collision coordinates[33b] for an A+BC system) is not useful in this case.

The situation is actually much worse for H-atom transfer in a polyatomic system, e.g., (3.1), than for the atom-diatom case (3.2), because the minimum energy path undergoes many sharp turns (in 3N-6 dimensional space) on its way from the transition state down to reactants and products. In fact one knows in general that the steepest descent path approaches a local minimum on the potential surface (i.e., reactants or products) along the normal mode of lowest frequency.[39] For reaction (3.1), for example, the steepest descent path begins at the saddle point being mostly motion of the H-atom that is transferred, but in moving downhill it switches successively to other motions, finally approaching the potential minimum along some in-plane skeletal vibration, the in-plane mode of lowest frequency. This "kinky" path is clearly not appropriate for defining a reaction coordinate.

To deal with H-atom transfer reactions in polyatomic systems, such as (3.1), we have previously suggested[39] using a straight-line Cartesian path[40] on which to base the dynamical model. The purpose of this paper is to develop this idea in a more rigorous fashion than before, correctly incorporating conservation of total angular (and, trivially, linear) momentum. We also show rigorously how all coupling in the kinetic energy part of the Hamiltonian can be eliminated, it then appearing in the potential energy. For this reason we have termed this model a diabatic reaction path Hamiltonian in analogy with the adiabatic/diabatic language used for describing systems with electronic and nuclear (i.e., vibration, rotation, translation) degrees of freedom.[41] Following this analogy, the original reaction path Hamiltonian[15] would be called the adiabatic reaction path Hamiltonian since the local vibrational modes orthogonal to the reaction path are the exact normal modes for a fixed value of the reaction coordinate (i.e., a fixed position on the reaction path); coupling between these modes and the reaction coordinate appears in the kinetic energy, just as does the coupling between nuclear degrees of freedom and adiabatic electronic states. In the model developed in this paper, coupling between the reaction coordinate and perpendicular modes has been transformed from the kinetic to the potential energy, the same as for a diabatic electronic representation.

It is useful to discuss qualitatively why we think a linear reference, or reaction path, will be useful for H-atom transfer reactions. Fig. 4a shows the sketch of (contours of) a potential energy surface typical of a collinear heavy + light-heavy system, e.g., reaction (3.2) It is well known[38,42] in such cases that the tunneling dynamics does not follow the minimum energy path (the full line) but rather "cuts-the-corner"; the linear path from reactants to products is the extreme version of this. Fig. 4a also pertains to certain modes in a polyatomic system that have a predominantly symmetric type of coupling; for example, the 2-d potential surface for reaction (3.1) that includes the reaction coordinate (s-coordinate) and the O-O stretch (Q-coordinate) looks qualitatively like Fig. 4a.

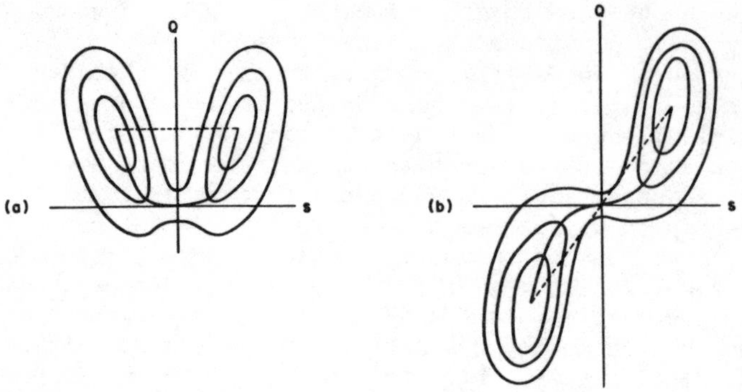

Figure 4.   Sketch of contour plots for two characteristic potential energy surfaces.   The solid lines indicate the minimum energy path from the transition state down to reactants and to products, and the broken line is the straight line path from reactants to products.

Fig. 4b, on the other hand, is for a mode with predominantly asymmetric coupling to the reaction coordinate, one for which the potential well in the reaction coordinate is asymmetric for a fixed (non-zero) value of the other coordinate.   The minimum energy path in this case will also be sharply curved and not useful for defining a reaction coordinate.   The straight-line path in this case "cuts" both corners, passing through the transition state.   An example of this situation is the double H-atom transfer in formic acid dimer,

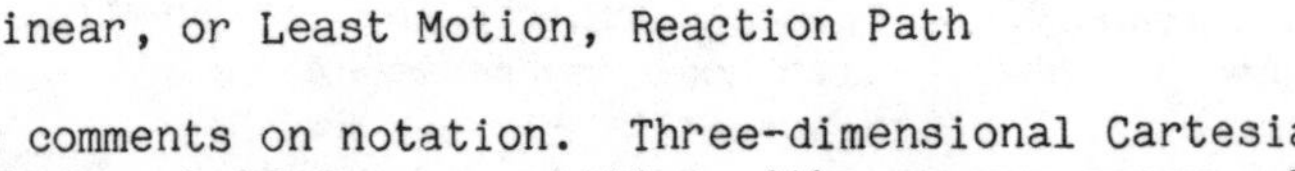

$$
\text{H}-\text{C}\begin{array}{c}\text{O}-\text{H}\cdots\text{O}\\ \diagdown\\ \text{O}\cdots\text{H}-\text{O}\end{array}\text{C}-\text{H} \quad\rightarrow\quad \text{H}-\text{C}\begin{array}{c}\text{O}\cdots\text{H}-\text{O}\\ \diagdown\\ \text{O}-\text{H}\cdots\text{O}\end{array}\text{C}-\text{H}, \qquad (3.3)
$$

where the coordinate s of Fig. 4b is the concerted motion of the two H atoms and Q the asymmetric O-C-O stretch that is coupled strongly to it.

3.2. The Linear, or Least Motion, Reaction Path

First some comments on notation.   Three-dimensional Cartesian vectors are indicated as bold-face quantities with an over arrow.   Thus $\vec{R}_i$, i=1,...,N, are the Cartesian coordinates of the N atoms; $\vec{x}_i$ are the corresponding mass-weighted coordinates

$$
\vec{x}_i = \sqrt{m_i}\ \vec{R}_i. \qquad (3.4a)
$$

Bold-face x with no index i is the 3N dimensional vector $\{x_{i\gamma}\}$, γ=x,y,z, i=1, ..., N.   We will switch on occasion between vector notation and component notation; thus in component notation Eq. (3.4a) is

$$x_{i\gamma} = \sqrt{m_i}\, R_{i\gamma}. \tag{3.4b}$$

The linear reaction path is defined by linear interpolation between reactant and product geometry, i.e.,

$$\underset{\sim}{x}_0(s) = \tfrac{1}{2}(\underset{\sim}{x}_r+\underset{\sim}{x}_p) + (\underset{\sim}{x}_p-\underset{\sim}{x}_r)(s/\Delta s) \tag{3.5a}$$

$$\Delta s = |\underset{\sim}{x}_p-\underset{\sim}{x}_r|, \tag{3.5b}$$

where $\underset{\sim}{x}_r$, $\underset{\sim}{x}_p$ $(\equiv\{x_{i\gamma}^{(r)}\},\{x_{i\gamma}^{(p)}\})$ are the 3N mass-weighted Cartesian coordinates of the atoms for the equilibrium geometry of the reactants and products, respectively. In terms of the coordinates $\vec{\underset{\sim}{R}}_i$, Eq. (3.5) is

$$\vec{\underset{\sim}{R}}_i^{(0)}(s) = \tfrac{1}{2}(\vec{\underset{\sim}{R}}_i^{(r)}+\vec{\underset{\sim}{R}}_i^{(p)}) + (\vec{\underset{\sim}{R}}_i^{(p)}-\vec{\underset{\sim}{R}}_i^{(r)})(s/\Delta s). \tag{3.5c}$$

s, the reaction coordinate, is the distance along this path, and as s varies from $-\Delta s/2$ to $+\Delta s/2$ the reference geometry varies from that of reactants to products. We note that

$$\underset{\sim}{x}_0'(s) = (\underset{\sim}{x}_p-\underset{\sim}{x}_r)/\Delta s, \tag{3.6a}$$

so that

$$|\underset{\sim}{x}_0'(s)| = 1, \tag{3.6b}$$

where prime denotes (d/ds).

To make the above definitions concrete we must specify how the axis system which defines product coordinates $\vec{\underset{\sim}{R}}_i^{(p)}$ is related to the axis that is used to define the reactant coordinates $\vec{\underset{\sim}{R}}_i^{(r)}$. This is intimately connected with the requirement[43] that the reference path $\underset{\sim}{x}_0(s)$ be one for which no linear or angular momentum be generated for displacements along it. I.e., to use the Hougen-Bunker-Johns[43] methodology the path $\underset{\sim}{x}_0(s)$ must satisfy the conditions

$$0 = \sum_i m_i\, \vec{\underset{\sim}{R}}_i^{(0)\prime}(s) \tag{3.7a}$$

$$0 = \sum_i m_i \, \vec{R}_i^{(0)}(s) \times \vec{R}_i^{(0)\prime}(s). \tag{3.7b}$$

Eq. (3.7a) (i.e., no linear momentum along the path) is easily satisfied by choosing the reactant and product coordinates so that the center of mass in each case is at the origin, i.e., so that

$$\sum_i m_i \, \vec{R}_i^{(r)} = \sum_i m_i \, \vec{R}_i^{(p)} = 0. \tag{3.8}$$

If the reactant coordinates, for example, are originally given so that Eq. (3.8) is not true, then one re-defines them by

$$\vec{R}_i^{(r)} \to \vec{R}_i^{(r)} - \vec{R}_{com}^{(r)}, \tag{3.9a}$$

where

$$\vec{R}_{com}^{(r)} = \sum_i m_i \, \vec{R}_i^{(r)} / \sum_i m_i . \tag{3.9b}$$

Multiplying Eq. (3.5c) by m_i and summing over i gives (using Eq. (3.8))

$$0 = \sum_i m_i \, \vec{R}_i^{(0)}(s), \tag{3.10}$$

and Eq. (3.7a) easily follows by differentiation.

Insuring no angular momentum along the linear reaction path is trickier. Substituting the linear path, Eq. (3.5c), into Eq. (3.7b) leads to the equation

$$0 = \sum_i m_i \left(\vec{R}_i^{(r)} \times \vec{R}_i^{(p)} \right) \tag{3.11}$$

as the condition that the total angular momentum remains zero along the path. Eq. (3.11) will not be true unless the product axis system is oriented in precisely the correct way with respect to the reactant one. (The two axis systems have already been chosen so that the center of mass of both is at the origin.) Thus suppose that Eq. (3.11) is not true for the initial orientation of the product axis system (with respect to the reactant one). One then rotates the product axis, i.e., the product coordinates are replaced by

$$\vec{R}_i^{(p)} \to \underset{\approx}{T} \cdot \vec{R}_i^{(p)}, \tag{3.12}$$

where $\underset{\approx}{T}$ is the 3x3 Cartesian rotation matrix[44] parameterized by three Euler angles that specify the rotation. These three Euler angles are chosen so that the three equations in Eq. (3.11) are satisfied.

It is useful to see explicitly how this works for the case that reactant and product molecule are planar, e.g., as for reactions (3.1) and (3.3). The reactant and product coordinate vectors thus have the form

$$\vec{\underset{\sim}{R}}_i^{(r)} = \begin{pmatrix} X_i^{(r)} \\ Y_i^{(r)} \\ 0 \end{pmatrix}, \quad \vec{\underset{\sim}{R}}_i^{(p)} = \begin{pmatrix} X_i^{(p)} \\ Y_i^{(p)} \\ 0 \end{pmatrix}, \tag{3.13}$$

and it is then easy to show that Eq. (3.11) reduces to the single equation

$$0 = \sum_i m_i \left(X_i^{(r)} Y_i^{(p)} - Y_i^{(r)} X_i^{(p)}\right) \equiv \sum_i m_i \left(\vec{\underset{\sim}{R}}_i^{(r)} \times \vec{\underset{\sim}{R}}_i^{(p)}\right)_z. \tag{3.14}$$

If Eq. (3.14) is not true, then the product axis system is rotated by an angle ϕ about the z-axis, whereby $\vec{\underset{\sim}{R}}_i^{(p)}$ of Eq. (3.13) is replaced by

$$\begin{pmatrix} X_i^{(p)} \\ Y_i^{(p)} \\ 0 \end{pmatrix} \rightarrow \begin{pmatrix} \cos\phi\ X_i^{(p)} + \sin\phi\ Y_i^{(p)} \\ -\sin\phi\ X_i^{(p)} + \cos\phi\ Y_i^{(p)} \\ 0 \end{pmatrix}. \tag{3.15}$$

With this replacement it is a simple calculation to show that (3.14) becomes

$$0 = -\sin\phi \sum_i m_i \left(X_i^{(r)} X_i^{(p)} + Y_i^{(r)} Y_i^{(p)}\right)$$
$$+ \cos\phi \sum_i m_i \left(X_i^{(r)} Y_i^{(p)} - Y_i^{(r)} X_i^{(p)}\right),$$

which is satisfied by the choice

$$\phi = \tan^{-1}\left[\frac{\sum_i m_i \left(\vec{\underset{\sim}{R}}_i^{(r)} \times \vec{\underset{\sim}{R}}_i^{(p)}\right)_z}{\sum_i m_i\ \vec{\underset{\sim}{R}}_i^{(r)} \cdot \vec{\underset{\sim}{R}}_i^{(p)}}\right]. \tag{3.16}$$

Thus if the original product coordinates $\vec{\underset{\sim}{R}}_i^{(p)}$ do not satisfy Eq. (3.14), they are rotated according to Eq. (3.15), with the angle ϕ given by Eq. (3.16).

The requirement of no linear and angular momentum along the

reaction path, Eq. (3.7), thus uniquely defines the axis system for the product coordinates with respect to that for the reactant.

3.3. Reaction Path Hamiltonian for a Linear Reaction Path

With the linear reaction path defined as in the previous section, one can proceed to construct the Hamiltonian in precisely the same manner as for the original reaction path Hamiltonian.[15] Thus the reaction path Hamiltonian for J=0 is given by

$$H\left(P_s,s,\{P_k,Q_k\}\right) = \tfrac{1}{2}\Big[P_s - \sum_{k,k'=1}^{3N-7} Q_k P_{k'} B_{k,k'}(s)\Big]^2$$

$$+ \sum_{k=1}^{3N-7} \tfrac{1}{2}P_k^{\;2} + V_0(s) - \sum_{k=1}^{3N-7} Q_k f_k(s) + \sum_{k=1}^{3N-7} \tfrac{1}{2}\omega_k(s)^2 Q_k^{\;2}, \quad (3.17)$$

where

$$f_k(s) = -\sum_{i\gamma} D_{i\gamma}(s) L_{i\gamma,k}(s) \quad (3.18a)$$

$$D_{i\gamma}(s) = \left(\frac{\partial V}{\partial x_{i\gamma}}\right)_{\tilde{x}=\tilde{x}_0(s)} \quad (3.18b)$$

and

$$B_{k,k'}(s) = \sum_{i\gamma} L_{i\gamma,k}{}'(s) L_{i\gamma,k'}(s). \quad (3.19)$$

In the above equations $\{L_{i\gamma,k}(s)\}$, k=1, ..., 3N-7 are, as before,[15] the eigenvectors of the projected force constant matrix along the reaction path, and $\{\omega_k(s)^2\}$ are the eigenvalues.

Eq. (3.17) is the same as the original reaction path Hamiltonian[15] with two exceptions. First, because the reaction path is straight, the curvature coupling elements $B_{k,3N-6}(s) \equiv 0$, so the first term in the Hamiltonian does not have the factor

$$\Big[1 + \sum_{k=1}^{3N-7} Q_k B_{k,3N-6}(s)\Big]^2 \quad (3.20)$$

that appears in the denominator of the previous result.[15] Second, since the present linear reaction path is not the minimum energy path, the potential energy has a term that is linear in the coordinates $\{Q_k\}$. We note also that cubic and quartic terms in coordinates $\{Q_k\}$ can readily be added to Eq. (3.17) if the third and fourth Cartesian derivatives of the potential are evaluated along the reaction path. The cubic term, for example, is

$$\frac{1}{6} \sum_{k,k',k''=1}^{3N-7} Q_k Q_{k'} Q_{k''} C_{kk'k''}(s), \qquad (3.21a)$$

where

$$C_{kk'k''}(s) = \sum_{i\gamma} \sum_{i'\gamma'} \sum_{i''\gamma''} \left(\frac{\partial^3 V}{\partial x_{i\gamma}\partial x_{i'\gamma'}\partial x_{i''\gamma''}}\right)_{\underset{\sim}{x} = \underset{\sim}{x}_0(s)}$$

$$\cdot L_{i\gamma,k}(s) L_{i'\gamma',k'}(s) L_{i''\gamma'',k''}(s), \qquad (3.21b)$$

and the quartic term is similar. It is, of course, possible to include such higher order terms in only some modes k and not in others. Finally, we note that the Hamiltonian for $J>0$ is also constructed in the present case in the same manner as before.[15]

3.4. Elimination of Kinetic Energy Coupling

The final step in obtaining the diabatic reaction path Hamiltonian is to eliminate the "coriolis" coupling terms in Eq. (3.17) which involve the coupling elements $B_{k,k'}(s)$. Since this procedure has been carried out before,[35a] the result is given here without derivation. The diabatic reaction path Hamiltonian is thus given by (using matrix notation)

$$H(P_s,s,\underset{\sim}{P},\underset{\sim}{Q}) = \tfrac{1}{2}P_s^2 + \tfrac{1}{2}\underset{\sim}{P}^T\cdot\underset{\sim}{P} + V_0(s) - \underset{\sim}{f}(s)^T\cdot\underset{\sim}{Q} + \tfrac{1}{2}\underset{\sim}{Q}^T\cdot\underset{\approx}{\Lambda}(s)\cdot\underset{\sim}{Q}, \qquad (3.22a)$$

with

$$\underset{\sim}{f}^T(s) = -\underset{\sim}{D}^T(s)\cdot\underset{\approx}{M}(s) \qquad (3.22b)$$

$$\underset{\approx}{\Lambda}(s) = \underset{\approx}{M}^T(s)\cdot\underset{\approx}{K}(s)\cdot\underset{\approx}{M}(s), \qquad (3.22c)$$

where the $3N\times(3N-7)$ transformation matrix $\underset{\approx}{M}(s)$ is

$$\underset{\approx}{M}(s) \equiv \underset{\approx}{L}(s)\cdot\underset{\approx}{U}(s). \qquad (3.23)$$

and where the matrix $\{U_{k,k'}(s)\}$, k, $k'=1, \ldots, 3N-7$, is defined by the equation

$$\underset{\approx}{U}'(s) = \underset{\approx}{B}(s)\cdot U(s). \qquad (3.24)$$

$\tilde{D}(s)$ and $\underset{\approx}{K}(s)$ are the Cartesian gradient and force constant matrix, and we have emphasized that it is only the combination $\underset{\approx}{L}(s)\cdot\underset{\approx}{U}(s) \equiv \underset{\approx}{M}(s) \equiv \{M_{i\gamma,k}(s)\}$ that is required to construct the quantities that go in the Hamiltonian. We will discuss below how this transformation matrix $\underset{\approx}{M}(s)$ is determined.

The procedure for constructing the Hamiltonian is thus as follows: First the linear reaction path is properly determined as in Section 3.2 from the reactant and product equilibrium geometries. One then computes the energy $V_0(s)$, Cartesian gradient $\tilde{D}(s)$, and Cartesian force constant matrix $\underset{\approx}{K}(s)$ along this path (and also higher derivatives of the potential, eq. Eq. (3.21), if these are desired). The transformation matrix $\underset{\approx}{M}(s)$ is then determined as below and the quantities $\tilde{f}(s)$ and $\underset{\approx}{\Lambda}(s)$ are computed via Eq. (3.22b) and (3.22c). If cubic, quartic, etc., terms in the potential are required, then the Cartesian terms, e.g., Eq. (3.21), are transformed from Cartesian space to Q_k-space via the matrix $\underset{\approx}{M}(s)$.

To conclude this section we show a simple procedure for determining the transformation matrix $\underset{\approx}{M}(s)$ of Eq. (3.23). (To make the notation below less cluttered we do not always denote the explicit s-dependence of the quantities $\underset{\approx}{L}$, $\underset{\approx}{M}$, and $\underset{\approx}{U}$.) By using the definition of the coupling matrix $\underset{\approx}{B}(s) \equiv \{\tilde{B}_{k,k'}(s)\}$,

$$\underset{\approx}{B}(s) = \underset{\approx}{L}^{T'}(s)\cdot\underset{\approx}{L}(s), \tag{3.25}$$

and the orthogonality and completeness relations of the matrix of eigenvectors $\underset{\approx}{L}(s) \equiv \{L_{i\gamma,k}(s)\}$,

$$\underset{\approx}{L}^{T}(s)\cdot\underset{\approx}{L}(s) = \underset{\approx}{1} \tag{3.26a}$$

$$\underset{\approx}{L}(s)\cdot\underset{\approx}{L}^{T}(s) = \underset{\approx}{1} - \underset{\approx}{P}(s), \tag{3.26b}$$

where $\underset{\approx}{P}(s) \equiv \{P_{i\gamma,i'\gamma'}(s)\}$ is the projector[15] onto the six directions that are overall translation and rotation of the N-atom system, one can derive the following first order differential equation for $\underset{\approx}{M}(s)$,

$$\underset{\approx}{M}'(s) = -\underset{\approx}{P}'(s)\cdot\underset{\approx}{M}(s) \tag{3.27}$$

which can now take as the fundamental defining equation for $\underset{\approx}{M}(s)$. One needs only to supplement it with a boundary (i.e., initial condition), e.g.,

$$\underset{\approx}{M}(0) = \underset{\approx}{L}(0)\cdot\underset{\approx}{U}(0);$$

if we choose $\underline{\underline{U}}(0) = \underline{\underline{1}}$, then the initial condition is

$$\underline{\underline{M}}(0) = \underline{\underline{L}}(0), \tag{3.28}$$

where $\underline{\underline{L}}(0)$ is obtained by diagonalizing the force constant matrix at the single position s = 0. With Eq. (3.28) as the initial condition for $\underline{\underline{M}}(s)$, the differential equation Eq. (3.27) determines it at all other values of s.

Integrating Eq. (3.27) over a short increment (s_{k-1}, s_k) gives

$$\underline{\underline{M}}_k - \underline{\underline{M}}_{k-1} = -(\underline{\underline{P}}_k - \underline{\underline{P}}_{k-1}) \cdot \underline{\underline{M}}_{k-1}, \tag{3.29}$$

where

$$\underline{\underline{M}}_k = \underline{\underline{M}}(s_k)$$

$$\underline{\underline{P}}_k = \underline{\underline{P}}(s_k),$$

etc. Since

$$\underline{\underline{P}}(s) \cdot \underline{\underline{M}}(s) = 0$$

for all s (because $\underline{\underline{P}}(s) \cdot \underline{\underline{L}}(s) = 0$), Eq. (3.29) becomes

$$\underline{\underline{M}}_k = (\underline{\underline{1}} - \underline{\underline{P}}_k) \cdot \underline{\underline{M}}_{k-1}. \tag{3.30}$$

Iterating this relation gives

$$\underline{\underline{M}}_k = (\underline{\underline{1}} - \underline{\underline{P}}_k) \cdot (\underline{\underline{1}} - \underline{\underline{P}}_{k-1}) \cdots (\underline{\underline{1}} - \underline{\underline{P}}_1) \cdot \underline{\underline{M}}_0 \tag{3.31}$$

as a simple way to compute $\underline{\underline{M}}$ over a grid of $\{s_k\}$ values, given the initial condition $\underline{\underline{M}}(0)$, i.e., Eq. (3.28).

In summary then, the matrix $\underline{\underline{M}}(s)$ that transforms from the Cartesian space (iγ) to the diabatic space (k) is given by Eq. (3.31), where the initial value $\underline{\underline{M}}(0) \equiv \underline{\underline{L}}(0)$, Eq. (3.28), is determined by diagonalizing the projected force constant matrix at the one position s = 0. It is not necessary to diagonalize the projected force constant matrix at any other values; only the projectors $\underline{\underline{P}}(s_k)$ are needed at the various values of the reaction coordinate in Eq. (3.31).

Finally, throughout Section 3 the use of <u>classical mechanics</u> has been implicitly assumed. Because the resulting Hamiltonian, Eq. (3.22), has a Cartesian kinetic energy, though, it is trivial to transform the result to a quantum mechanical Hamiltonian operator; i.e., in Eq. (3.22) one makes the standard replacements

$$\tfrac{1}{2}\, P_s^{\,2} \rightarrow -\frac{\hbar^2}{2}\frac{\partial^2}{\partial s^2}$$

$$\tfrac{1}{2}\, P_k^{\,2} \rightarrow -\frac{\hbar^2}{2}\frac{\partial^2}{\partial Q_k^{\,2}}.$$

4. CONCLUDING REMARKS

It has been a pleasure to present this work to a group consisting largely of quantum chemists, for I believe that both topics are quite timely for this audience. First, the approach to reactive scattering is seen to reduce to quite standard quantum mechanics: choosing basis functions, computing matrix elements of the Hamiltonian, and then performing a large linear algebra calculation. Because quantum chemists have so much experience and have developed sophisticated methodologies for carrying out these tasks in electronic structure calculations, I believe that much of their expertise can now be fruitfully applied to reactive scattering.

Second, the new diabatic reaction path Hamiltonian gives one a framework for using *ab initio* quantum chemistry calculations to treat a new class of dynamical processes in polyatomic molecules. It is actually much simpler to apply than the original version, based on the minimum energy path, because for the new diabatic version one does not need to generate the minimum energy path. I.e., one needs to determine only the reactant and product geometries and then compute the energy, gradient, and force constant matrix (and higher derivatives if desired) along a *pre-determined* (i.e., the linear interpolation) path. The form of the diabatic reaction path Hamiltonian, having a Cartesian kinetic energy, is also much simpler for purposes of carrying out dynamics calculations. It should be especially useful for describing H-atom transfer reactions in polyatomic systems.

ACKNOWLEDGMENTS

I would like to acknowledge some very stimulating discussions with Professor C. W. McCurdy on the possibility of using the complex scaling/coordinate rotation approach to calculate $G^+(E)$, and thus $T = V+VG^+V$, that led ultimately to the present S-matrix version of the Kohn variational principle. I also gratefully acknowledge generous support of this work by the Science Foundation, grant CHE84-16345, and by the Director, Office of Energy Research, Office of Basic Energy Sciences, Chemical Sciences Division of the U.S. Department of Energy, under Contract Number DE-AC03-76SF00098.

REFERENCES

1. W. H. Miller, J. Chem. Phys. 50, 407 (1969).
2. There are several approaches to reactive scattering that do not involve non-local interactions: the hyperspheric coordinate method, see for example, (a) A. Kuppermann and P. G. Hipes, J. Chem. Phys. 84, 5962 (1986), and (b) C. A. Parker, R. T. Pack, B. J. Archer, and R. B. Walker, Chem. Phys. Lett. 137, 564 (1987); the matching-on-a-surface-method, (c) A. Kuppermann and G. C. Schatz, J. Chem. Phys. 62, 2502 (1975); and the use of a reaction coordinate, see for example, (d) M. J. Redmon and R. E. Wyatt, Chem. Phys. Lett. 63, 209 (1979), and (e) R. B. Walker, E. B. Stechel, and J. C. Light, J. Chem. Phys. 69, 2922 (1978).
3. Attempts at solving the equations by an iterative (SCF-like) scheme failed to coverage. W. H. Miller, unpublished results 1969.
4. See, for example, T.-Y. Wu and T. Ohmura, Quantum Theory of Scattering, (Prentice-Hall, Englewood Cliffs, New Jersey, 1962), p. 57-68.
5. W. Kohn, Phys. Rev. 74, 1763 (1948). See also, ref. 4.
6. (a) C. Schwartz, Phys. Rev. 124, 1468 (1961).
   (b) Ann. Phys. (New York) 10, 36 (1961).
7. R. K. Nesbet, Variational Methods in Electron-Atom Scattering Theory, (Plenum, New York, 1980), pp. 30-50.
8. (a) W. H. Miller and B. M. D. D. Jansen op de Haar, J. Chem. Phys. 86, 6213 (1987).
   (b) J. Z. H. Zhang, S.-I. Chu, and W. H. Miller, J. Chem. Phys. 88, 6233 (1988).
   (c) J. Z. H. Zhang and W. H. Miller, Chem. Phys. Lett. 140, 329 (1987).
   (d) J. Chem. Phys. 88, 4549 (1988).
9. C. W. McCurdy, T. N. Rescigno, and B. I. Schneider, Phys. Rev. A 36, 2061 (1987).
10. (a) K. Haug, D. W. Schwenke, Y. Shima, D. G. Truhlar, J. Z. H. Zhang, and D. J. Kouri, J. Phys. Chem. 90, 6757 (1986); 92, 3202 (1988).
    (b) J. Z. H. Zhang, D. J. Kouri, K. Haug, D. W. Schwenke, Y. Shima, and D. G. Truhlar, J. Chem. Phys. 88, 0000 (1988).
11. M. Baer and D. J. Kouri, Phys. Rev. A 4, 1924 (1971); for a recent review, see D. J. Kouri, in Theory of Chemical Reaction Dynamics, ed. M. Baer, (CRC Press, Boca Raton, Florida, 1985), p. 163.
12. R. G. Newton, Scattering Theory of Waves and Particles, (McGraw-Hill, New York, 1966), Sec. 11.3.
13. B. R. Johnson and D. Secrest, J. Math. Phys. 7, 2187 (1966).

14. W. H. Miller, B. A. Ruf, and Y.-T., Chang, J. Chem. Phys. 89, 0000 (1988).
15. W. H. Miller, N. C. Handy, and J. E. Adams, J. Chem. Phys. **72**, 99 (1980).
16. A. D. Isaacson, C. W. McCurdy, and W. H. Miller, Chem. Phys. 34, 311 (1978).
17. A. J. F. Siegert, Phys. Rev. 56, 750 (1939).
18. Ref. 7, p. 35 et seq.
19. (a) J. Nuttall and H. L. Cohen, Phys. Rev. 188, 1542 (1969).
(b) F. A. McDonald and J. Nuttall, Phys. Rev. C 6, 121 (1972).
(c) R. T. Baumel, M. C. Crocker, and J. Nuttall, Phys. Rev. A 12, 486 (1975).
20. (a) T. N. Rescigno and W. P. Reinhardt, Phys. Rev. A 8, 2828 (1973); 10, 158 (1974).
(b) B. R. Johnson and W. P. Reinhardt, ibid. 29, 2933 (1984).
21. C. W. McCurdy and T. N. Rescigno, Phys. Rev. A 21, 1499 (1980); 31, 624 (1985).
22. For reviews, see
(a) B. R. Junker, Adv. At. Mol. Phys. 18, 207 (1982).
(b) W. P. Reinhardt, Annu. Rev. Phys. Chem. 33, 223 (1982).
(c) Y. K. Ho, Phys. Rep. 99, 1 (1983).
23. G. C. Schatz, J. M. Bowman, and A. Kuppermann, J. Chem. Phys. 63, 674 (1975).
24. (a) J. M. Bowman, Adv. Chem. Phys. 61, 115 (1985).
(b) Chem. Phys. Lett. 141, 545 (1987).
25. J. V. Lill, G. A. Parker, and J. C Light, J. Chem. Phys. 85, 900 (1986).
26. R. A. Friesner, Chem. Phys. Lett. 116, 39 (1985).
27. See, for example, H. F. Schaefer, The Electronic Structure of Atoms and Molecules, (Addison-Wesley, Reading Massachusetts, 1972), p. 70.
28. T. N. Rescigno and B. I. Schneider, Phys. Rev. A 37, 1044 (1988).
29. J. Z. H. Zhang and W. H. Miller, J. Chem. Phys. 89, 0000 (1988).
30. (a) L. M. Hubbard, S.-H. Shi, and W. H. Miller, J. Chem. Phys. 78, 2381 (1983).
(b) G. C. Schatz, L. M. Hubbard, P. S. Dardi, and W. H. Miller, J. Chem. Phys. 81, 231 (1984).
(c) P. S. Dardi, S.-H. Shi, and W. H. Miller, J. Chem. Phys. 83, 575 (1985).
31. J. Z. H. Zhang and W. H. Miller, to be published.
32. For recent applications, see A. Nauts and R. E. Wyatt, Phys. Rev. Lett. 51, 2238 (1983); Phys. Rev. A 30, 872 (1984); R. Friesner and R. E. Wyatt, J. Chem. Phys. 82, 1973 (1985).
33. For early work on reaction paths and reaction coordinates, see
(a) S. Glasstone, K. J. Laidler, and H. Eyring, The Theory of Rate Processes, McGraw-Hill, New York, 1941.
(b) R. A. Marcus, J. Chem. Phys. 45, 4493, 4500 (1966); 49, 2610 (1968).
(c) G. L. Hofacker, Z. Naturforsch. Teil A 18, 607 (1963); J. Chem. Phys. 43, 208 (1965).

(d) S. F. Fischer, G. L. Hofacker, and R. Seiler, J. Chem. Phys. 51, 3941 (1969).

34. Some papers by other workers on reaction path models are
(a) S. F. Fischer and M. A. Ratner, J. Chem. Phys. 57, 2769 (1972).
(b) P. Russegger and J. Brickman, ibid. 62, 1086 (1975); 60, 1 (1977).
(c) M. V. Basilevsky, Chem. Phys. 24, 81 (1977); 67, 337 (1982); M. V. Basilevsky and A. G. Shamov, ibid. 60, 349 (1981).
(d) K. Fukui, S. Kato, and H. Fujimoto, J. Am. Chem. Soc. 97, 1 (1975); K. Yamashita, T. Yamabe, and K. Fukui, Chem. Phys. Lett. 84, 123 (1981); A. K. Fukui, Acc. Chem. Res. 14, 363 (1981).
(e) K. Ishida, K. Morokuma, and A. Komornicki, J. Chem. Phys. 66, 2153 (1977).
(f) A. Nauts and X. Chapuisat, Chem. Phys. Lett. 85, 212 (1982); X. Chapuisat, A. Nauts, and G. Durrand, Chem. Phys. 56, 91 (1981).
(g) J. Pancir, Collect. Czech. Commun. 40, 1112 (1975); 42, 16 (1977).
(h) G. A. Natanson, Mol. Phys. 46, 481 (1982).
(i) M. Page and J. W. McIver, Jr., J. Chem. Phys. 88, 922 (1988).
(j) T. H. Dunning, Jr., and L. Harding, Faraday Disc. Chem. Soc., in press (1988).
(k) B. C. Garrett, M. J. Redmon, R. Steckler, D. G. Truhlar, K. K. Baldridge, D. Bartol, M. W. Schmidt, and M. S. Gordon, J. Phys. Chem. 92, 1476 (1988).

35. (a) W. H. Miller, in Potential Energy Surfaces and Dynamics Calculations, edited by D. G. Truhlar, Plenum, New York, 1981, p. 265.
(b) C. J. Cerjan, S.-H. Shi, and W. H. Miller, J. Phys. Chem. 86, 2244 (1982).
(c) W. H. Miller, J. Phys. Chem. 87, 3811 (1983).
(d) S. K. Gray, W. H. Miller, Y. Yamaguchi, and H. F. Schaefer, J. Chem. Phys. 73, 2733 (1980).
(e) S. K. Gray, W. H. Miller, Y. Yamaguchi, and H. F. Schaefer, J. Am. Chem. Soc. 103, 1900 (1981).
(f) Y. Osamura, H. F. Schaefer, S. K. Gray, and W. H. Miller, J. Am. Chem. Soc. 103, 1904 (1981).
(g) B. A. Waite, S. K. Gray, and W. H. Miller, J. Chem. Phys. 78, 259 (1983).
(h) W. H. Miller and S. Shi, J. Chem. Phys. 75, 2258 (1981).
(i) W. H. Miller and S. Schwartz, J. Chem. Phys. 77, 2378 (1982).
(j) S. Schwartz and W. H. Miller, J. Chem. Phys. 79, 3759 (1983).
(k) T. Carrington, Jr., L. M. Hubbard, H. F. Schaefer, and W. H. Miller, J. Chem. Phys. 80, 4347 (1984).
(ℓ) T. Carrington, Jr. and W. H. Miller, J. Chem. Phys. 81, 3942 (1984).
(m) W. H. Miller, in The Theory of Chemical Reaction Dynamics, ed. D. C. Clary, Reidel, Boston, 1986, p. 27-45.

(n) T. Carrington, Jr., and W. H. Miller, J. Chem. Phys. 84, 4364 (1986).
(o) W. H. Miller, in Tunneling, eds. J. Jortner and B. Pullman, D. Reidel, Boston, 1986, pp. 91-101.

36. See also,
(a) R. T. Skodje, D. G. Truhlar, and B. C. Garrett, J. Phys. Chem. 85, 3019 (1981).
(b) R. T. Skodje, D. G. Truhlar, and B. C. Garrett, J. Chem. Phys. 77, 5955 (1982).
(c) A. D. Isaacson and D. G. Truhlar, ibid. 76, 1380 (1982).
(d) D. G. Truhlar, N. J. Kilpatrick, and B. C. Garrett, ibid. 78, 2438 (1983).
(e) R. T. Skodje and D. G. Truhlar, ibid. 79, 4882 (1983).
(f) R. T. Skodje, D. W. Schwenke, D. G. Truhlar, and B. C. Garrett, J. Phys. Chem. 88, 628 (1984).
(g) B. C. Garrett and D. G. Truhlar, J. Chem. Phys. 81, 309 (1984).

37. For a theoretical treatment, see
(a) J. Bicerano, H. F. Schaefer and W. H. Miller, J. Am. Chem. Soc. 105, 2550 (1983).
(b) Ref. 35(n).

38. See, for example,
(a) J. Manz and J. Römelt, Chem. Phys. Lett. 81, 179 (1981).
(b) J. A. Kaye and A. Kuppermann, ibid. 77, 573 (1981); 78, 546 (1981).
(c) V. K. Babamov and R. A. Marcus, J. Chem. Phys. 74, 1790 (1981).
(d) C. Hiller, J. Manz, W. H. Miller, and J. Römelt, ibid. 78, 3850 (1983).
(e) B. C. Garrett, D. G. Truhlar, A. F. Wagner, and T. H. Dunning, Jr., ibid. 45, 120 (1973).

39. B. A. Ruf and W. H. Miller, Faraday Disc. Chem. Soc., in press (1988).
40. A straight-line reaction path has also been used in a particular application by J. M. Bowman, K. T. Lee, H. Romanowski, and L. B. Harding, in Resonances, ed. D. G. Truhlar, ACS Symp. Series 263, 43 (1984).
41. See, for example, F. T. Smith, Phys. Rev. 179, 111 (1969).
42. See, for example, T. F. George and W. H. Miller, J. Chem. Phys. 56, 5722 (1972); 57, 2458 (1972).
43. J. T. Hougen, P. R. Bunker, and J. W. C. Johns, J. Mol. Spect. 34, 136 (1970).
44. See, for example, M. E. Rose, in Elementary Theory of Angular Momentum, Wiley, New York, 1957, p. 65.

CHARGE TRANSFER REACTION DYNAMICS IN SOLUTIONS

JAMES T. HYNES
Department of Chemistry and Biochemistry
University of Colorado
Boulder, CO 80309-0215 USA

ABSTRACT. In the conventional Transition State Theory of the rate of heavy particle charge transfer reactions in polar solvents, it is assumed that the solvent is always in equilibrium at each stage of the reaction coordinate for the reacting system. We discuss the breakdown of this assumption for a model S_N2 reaction in water via analytic theory and Molecular Dynamics computer simulation. We also discuss a theory for the role of the solvent in intramolecular proton transfers.

I. Introduction

It is well known that a polar solvent can exert a strong influence on the rate of chemical reactions involving the displacement of change (1). In this paper, we will outline some recent theoretical efforts designed to understand this role at a molecular level for two reaction classes: S_N2 nucleophilic displacement reactions $X^- + RY \rightarrow XR + Y^-$ and proton transfers $AH^+ + B \rightarrow A + H^+B$. It is worth stressing at the outset that each problem area requires significant input from quantum chemistry for progress to be made.

II. Nonequilibrium Solvation

The influence of a polar solvent on chemical reaction rates has been traditionally described in terms of equilibrium free energy considerations. Within the context of the standard Transition State Theory (TST) approach to these questions, one looks to the solvent effect on the activation free energy $\Delta G^{\neq}$ to interpret the variation of a reaction rate with solvents. This approach is for example often applied to heavy particle charge transfer reactions such as S_N1 dissociations $RX \rightarrow R^+ + X^-$ and S_N2 substitution reactions $X^- + RY \rightarrow XR + Y^-$. Fig. 1 illustrates a so-called potential of mean force for an S_N2 reaction giving the free energy of the system subject to this equilibrium assumption.

J. Jortner and B. Pullman (eds.), Perspectives in Quantum Chemistry, 83–95.

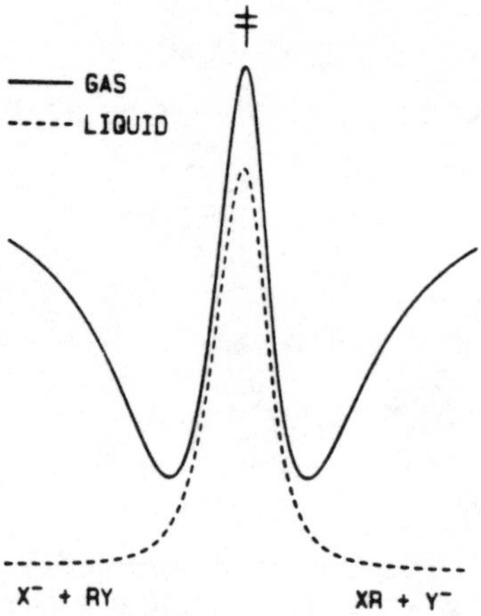

Figure 1. Schematic mean potential or free energy for an S_N2 reaction (lower curve). Gas phase potential showing ion dipole complex wells flanking a central barrier is also shown. The former results from the latter when a polar solvent is equilibrated to the reaction system at each value of the reaction coordinate.

Despite its success, this TST approach makes a basic assumption, namely, that of equilibrium solvation: the solvent is assumed to be always equilibrated at each stage along the reaction system's reaction coordinate. Thus the TST expression k^{TST} for the reaction rate constant k is independent of any solvent dynamics.

Clearly there is no guarantee that the TST equilibrium solvation assumption holds. For example, there can be insufficient time for solvent molecules to reorient to keep up with the reaction system's changing charge distribution as that system crosses an activation barrier. In such a case, there will be nonequilibrium solvation and k will differ from its TST value k^{TST}.

As shown by van der Zwan and Hynes for charge transfer reaction model studies (2), nonequilibrium solvation leads to a solvent induced recrossing of the barrier top -- an effect completely absent in the standard TST approach. The consequent reduction of k below its TST predicted value is measured by the transmission coefficient

$$\kappa = k/k^{TST}; \ \kappa \leq 1. \tag{1}$$

From a theoretical perspective, κ can be determined from Grote-Hynes (GH) theory (3, 4), which assumes the validity of a generalized Langevin equation (GLE) for the reaction coordinate x:

$$\ddot{x}(t) = \omega^2_{b,eq} x(t) - \int_0^t d\tau \zeta(t-\tau)\, \dot{x}(\tau), \qquad (2)$$

where we have ignored a random force term that is irrelevant in the present discussion. The first term in this GLE represents equilibrium solvation, with $\omega^2_{b,eq}$ being the square equilibrated barrier frequency, proportional to the absolute curvature of the solvated barrier (as e.g. in Fig. 1). The second contribution to the GLE represents the effect of nonequilibrium solvation. It involves the time dependent friction ζ (t), which is proportional to the time correlation function of the force that the solvent exerts on the reaction system coordinate x in the neighborhood of the barrier top or transition state. With the GLE eq. (2), the transmission coefficient is given by the self-consistent GH equation:

$$\kappa = \left(\kappa + \int_0^\infty dt\; e^{-\omega_{b,eq}\kappa t} \zeta(t)/\omega_{b,eq}\right)^{-1} . \qquad (3)$$

van der Zwan and Hynes (2) have found two important regimes in their analysis of GH theory in model studies of heavy particle charge transfer in solution. The first of these, the nonadiabatic regime, applies when barrier crossing can occur even when there is no solvent motion. Here κ is obtained by ignoring the time dependence of $\zeta(t)$ in the GH Eq. (3) and is given by (2)

$$\kappa_{NA} = \frac{\omega_{b,na}}{\omega_{b,eq}}; \qquad \omega^2_{b,na} = \omega^2_{b,eq} - \zeta(t=0) . \qquad (4)$$

This limit is most applicable for a "slowly" relaxing solvent, modest coupling of the reaction system to the solvent, and a large value of $\omega_{b,eq}$, i.e., a "sharp" barrier. The nonadiabatic frequency $\omega_{b,na}$ describes the curvature of an effective, nonequilibrated barrier that the reaction system experiences in its short time motion. $\kappa_{NA}<1$ is independent of any solvent dynamics so that the barrier crossing dynamics needs to be understood in terms of static solvent configurations rather than solvent dynamics.

The second regime is that of polarization caging (2), most likely for slowly relaxing solvents, strong coupling to the solvent and low $\omega_{b,eq}$ values. In this regime, the reaction system experiences a confining well or "polarization cage" in its short time motion; confining forces from the solvent dominate the chemical forces favoring passage off the barrier top towards products. On a longer time scale, the solvent motion destroys the cage, allowing the reaction to proceed. The key point here is that solvent motion is necessary for the reaction to occur; κ reflects this dependence on the solvent dynamics via explicit details of the time dependent friction in Eq. (3). The GH theory and these ideas can be tested by Molecular Dynamics (MD) computer simulation as now described.

II. S_N2 Reactions

MD simulations of a model of the symmetric S_N2 reaction $Cl^- + CH_3Cl \rightarrow ClCH_3 + Cl^-$ in H_2O have been carried out (5,6) to test these ideas. These studies used in part the gas phase ab initio quantum one dimensional potential generated by Jorgensen and coworkers (7). The three dimensional gas phase energy surface is of a LEPS form, with a central barrier of height 13.9 kcal/mol with respect to two ion-dipole complex minima. The solvent-solute interactions are Lennard-Jones potentials and Coulomb interactions with varying charges on the solute atoms. The flexible Watts model for the water molecules was used. A key feature of the reaction is the rapid charge switching between the attacking and leaving chlorine in the transition state neighborhood calculated in the ab initio study (7). This leads to a strong electrical coupling to the solvent water molecules.

Trajectories are initiated at the transition state (zero asymmetric stretch) using the Keck-Anderson-Bennett technique (8), and trajectories are followed forwards and backwards in time to ensure that trajectories have originated as reactants and terminate as products. The transmission coefficient κ is determined (5, 6) both by the Stable States Picture method of Hynes and coworkers (3, 9) and the reactive flux formulation of Chandler (10), which give equivalent results for this system.

Significant recrossing of the barrier top is observed in the simulations and a marked departure from TST is found: $\kappa_{MD} = 0.54 \pm 0.05$. It is also found that the fate of a trajectory whether reactive or not is very quickly decided in a time of order 0.02 ps or less. On this time scale there is essentially negligible important motion of the solvent molecules and the reaction is approximately in the nonadiabatic regime described above. This is the antithesis of the equilibrium solvation picture where the solvent responds rapidly to solvate the moving reaction system.

This short time scale suggests that the solvent-induced recrossings can be comprehended in terms of static solvent configurations when the reaction system is at the transition state. Indeed a detailed analysis confirms this (5). For example those trajectories initially heading towards products and which quickly recross to form reactants have an asymmetric solvation in which the solvent preferentially solvates the "left-hand" chlorine in $Cl^{\delta-}CH_3^{\delta+}Cl^{\delta-}$. This bias induces a recrossing when the attempt to form products, i.e., have the "right-hand" chloride be the leaving ion, is made. In contrast, for trajectories that cross the barrier without any subsequent recrossing, there is no solvent bias.

The simple nonadiabatic limit eq. (4) gives a good description of the MD results. Table I illustrates this for several choices of the rate of charge switching between the attacking and leaving chloride ions. In this table, a value of unity for the charge switching rate denotes the value thought to represent the actual $Cl^- + CH_3$ Cl system, and the other values listed indicate a ratio. Thus the rate at which the electronic charge distribution shifts in response to the motion of the heavy reacting system particles has a pronounced effect on the transmission coefficient.

Table I. Transmission coefficients for the 13.9 kcal mol^{-1} barrier Cl^- + CH_3Cl reaction in water.[a)]

switching rate	κ_{MD}	κ_{GH}	κ_{NA}	κ_{KR}
1/2	0.71	0.74	0.69	0.03
1	0.54	0.57	0.48	0.02
2	0.40	0.37	0.25	0.004

a) Error bars are approximately $\pm$ 0.05 for κ_{MD} and κ_{NA}, $\pm$ 0.10 for κ_{GH}, while κ_{KR} is only estimated to within a factor of 2.

The full GH theory prediction eq. (3) can be tested by evaluating the time dependent friction $\zeta(t)$. This is done by fixing the asymmetric stretch reaction coordinate x at its transition state value x = 0 and computing the time correlation function of the force along x. Fig. 2 shows that on the characteristic time scale associated with the barrier crossing there is little solvent dynamical evolution apparent in $\zeta(t)$.

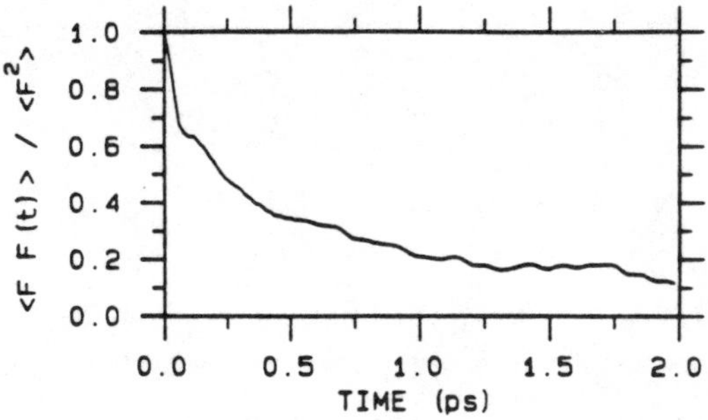

Figure 2. Normalized force correlation function at the transition state for the Cl^- + CH_3Cl reaction in H_2O.

Table I shows that GH theory is in excellent agreement with the MD simulation results. The improvement over the simple κ_{NA} prediction is due to the fact that some solvent motion influences the barrier recrossing, even though the systems are close to the nonadiabatic limit. A detailed analysis (6) shows that the most important solvent motions are the water librational motions. These motions are fast enough to exert an influence on the reaction time scale and are also strongly coupled to the reaction coordinate because they modulate the electrostatic solvent-solute interactions. In contrast, the fast H_2O stretching and bending vibrations exhibit weak coupling to the reaction coordinate.

Table I also displays the Kramers theory (11) transmission coefficients κ_{KR}, obtained from the GH eq. (3) by ignoring the "frequency dependence" of the friction and inserting the friction constant

$$\zeta = \int_0^\infty dt\, \zeta(t) .$$

The Kramers description is extremely poor. The reason for this is that the long time, "zero frequency" friction constant is quite large, predicting extensive barrier recrossing of a diffusive character. But these long time effects, which include the contributions of the "slow" water molecule reorientations for example, are in fact irrelevant on the short reaction time scale. It is this feature that accounts for the failure of Kramers theory and the success of GH theory.

The polarization cage limit can be reached in the simulations (6) by lowering the central barrier height so that the barrier frequency is diminished and by altering some details of the charge switching to increase the strength of the coupling to the solvent. The results for κ are shown in Table II.

Table II. Transmission coefficients for the 4.9 kcal mol^{-1} barrier S_N2 reaction in water[a].

switching rate	κ_{MD}	κ_{GH}	κ_{KR}
1/2	0.39	0.34	0.02
1	0.37	0.32	0.007
2	0.31	0.30	0.01

[a]Error bars are similar to those reported in Table I.

For these cases, the reaction system turns out to be just barely within the polarization caging regime. Inspection of the trajectories reveals that in fact the reaction system initially experiences a confining solvent cage on the average as it leaves the transition state as predicted. However rapid water librational motions rapidly destroy this cage before any extensive amount of barrier recrossing occurs. It can be seen from Table II that once again the GH theory gives an excellent account of the MD results, while Kramers theory is quite inadequate.

It is worth mentioning in passing that the Kramers description can sometimes apply when the barrier is quite low with a correspondingly low frequency. In this case, there is enough time for the full zero frequency friction constant to be relevant and GH theory reduces to Kramers theory. An example is furnished in an MD simulation of a solvent-separated ion pair → contact ion pair interconversion in a polar solvent (12). There it is found that $\kappa_{MD} = 0.2$, in good agreement with both κ_{GH} and κ_{KR} for this system.

We have seen in these S_N2 model studies that the role of the solvent on a molecular level is quite different from that usually envisaged, and that we have a theory that successfully describes that role. It is likely that this will also be true for other reaction classes in solution. In examining these questions though, it will be quite important, just as it was for the S_N2 case, to have reasonable quantum chemical input for the reaction potential surfaces, and for a description of the electronic charge redistribution within the reaction system.

III. Proton Transfers in Solution

We now turn to a reaction class of a different type: proton transfers of the form $AH^+ \cdots B \rightarrow A \cdots H^+B$ where the dashed lines indicate a hydrogen bond.

Three features which make proton transfer reactions special must be included in any reasonable description. First, the H^+ motion will be strongly coupled to intramolecular vibrational modes in the reaction system. For example, the barrier for H^+ transfer is found to be strongly dependent on e.g. O-O separations in quantum chemical calculations of $OH^+ \cdots O$ transfer systems. Second, the proton can strongly couple electrically to the surrounding polar solvent molecules which distort the H^+ transfer potentials. Illustrations of both of these features are provided in a number of ab initio calculations (13). Third, the light proton is a fundamentally quantum particle and can tunnel through a potential barrier rather than be activated over it.

Here we sketch some central features of a theory due to Borgis and Hynes (14) (BH) incorporating the aspects above for intramolecular proton transfers in polar solvents. This BH theory has some features in common with earlier pioneering studies of proton transfers (15); similarities and differences are discussed elsewhere (14). Here we focus on the key ideas and predictions of the BH theory.

The first important coordinate for an intramolecular proton transfer is q, the proton's coordinate. There is a double well potential in q which is illustrated in Fig. 3 for the schematic symmetric transfer $AH^+ \cdots A \rightarrow A \cdots H^+A$, where the A's represent heavy atoms such as O, C, N etc. The second important coordinate $Q = X - X_{eq}$ is related to the intramolecular separation between the A's. The Q dependence of the proton barrier can be very strong, particularly when there is a hydrogen bond (Fig. 3). In particular, the barrier height can plummet, at the rate (16) of ~ 50 kcal/mol/Å, when Q decreases. This feature can lead to a strong dynamical modulation of the tunneling rate as Q vibration occurs; the tunneling probability increases exponentially as the barrier height and width decrease (Fig. 3). We term the fluctuations in Q the "coupling fluctuations" since they modulate the coupling C, which is proportional to the square root of the tunneling probability permitting the tunneling. These fluctuations will assist the tunneling.

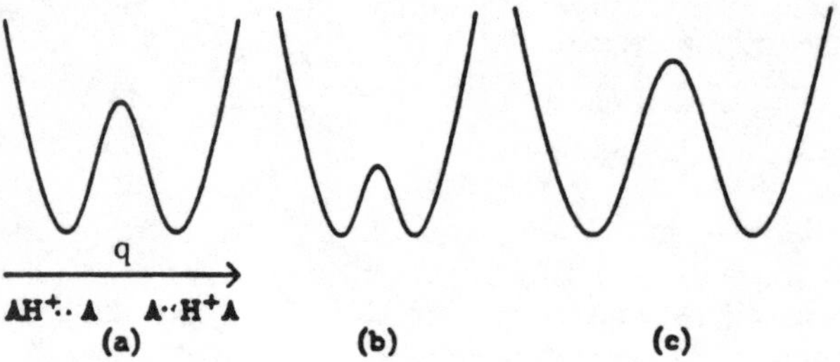

Figure 3. Proton double wells for (a) zero, (b) negative, and (c) positive values of the intramolecular coordinate $Q = X - X_{eq}$.

The third important coordinate is a collective solvent coordinate S, which we can think of for present purposes as measuring the solvent orientational polarization, i.e., the alignment of the polar solvent molecules. This coordinate will couple electrically to the charged proton coordinate q and will, in the first instance, modulate the symmetry of the proton transfer double well. If for example the solvent is equilibrated to, i.e. has the appropriate polarization for, the reactant $AH^+ \cdots A$, the left-hand well and the proton vibrational level (Fig.4) will be lowered while the right hand well and H^+ vibrational level will be raised in energy, introducing an asymmetry. If instead the solvent is equilibrated to $A \cdots H^+A$, then an asymmetry in the opposite sense exists.

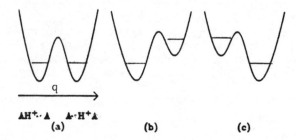

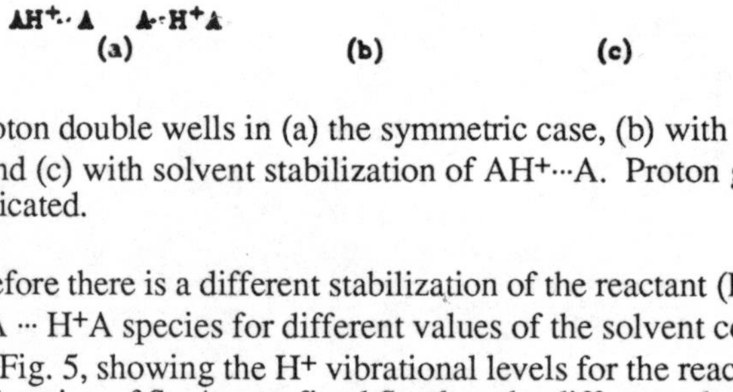

Figure 4. Proton double wells in (a) the symmetric case, (b) with solvent stabilization of $A \cdots H^+A$ and (c) with solvent stabilization of $AH^+ \cdots A$. Proton ground vibrational levels are indicated.

Therefore there is a different stabilization of the reactant (R) $AH^+ \cdots A$ and product (P) $A \cdots H^+A$ species for different values of the solvent coordinate S. This is illustrated in Fig. 5, showing the H^+ vibrational levels for the reactant and product species as a function of S. At any fixed S value, the difference between the two curves is the splitting $\Delta\varepsilon$, which equals the difference between the H^+ vibrational energies in the P and R species.

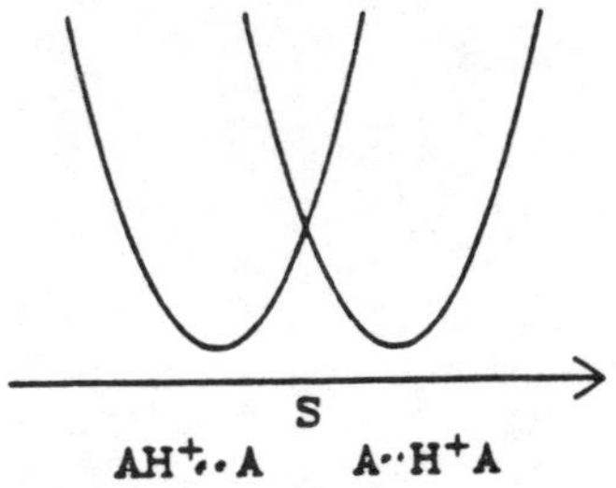

Figure 5. The proton energies for the reactant and product configurations as a function of the solvent coordinate S.

Motion in the solvent coordinate is thus important in modulating the symmetry of the q double well. This in turn strongly influences tunneling probabilities which increase rapidly with decreasing asymmetry, i.e. decreasing magnitude $|\Delta\varepsilon|$ of the splitting. Indeed these splitting fluctuations are required for tunneling to occur. Thus as solvent molecules e.g. rotate and reorient, there is motion in S in e.g. the left-hand well if the proton is initially in the $AH^+ \cdots A$ R configuration. Tunneling will not take place until the neighborhood of the intersection $S^{\neq}$ of the reactant and product curves in Fig. 5 is reached where the splitting $\Delta\varepsilon = 0$. At $S^{\neq}$, symmetry for the proton double well in Fig. 4 obtains, and tunneling can occur. Subsequent solvent motion then causes asymmetry and traps the proton in either R or P configurations. Attaining $S^{\neq}$ costs free energy, so the solvent will slow down the tunneling.

To evaluate the rate constant k for the H^+ transfer, BH use the time correlation function formula (17)

$$k = \int_0^\infty dt \;\; \mathrm{Re} < j(0)\, j(t) >, \tag{5}$$

in which j(0) is the initial $t = 0$ value of the proton probability flux out of the reactant $AH^+ \cdots A$ well into the product $A \cdots H^+A$ well and j(t) is its value at time t. The brackets indicate an equilibrium average over both the proton transfer system in its reactant state and the solvent, while Re denotes the real part. The greater is the flux correlation, the greater is its time integral, the rate constant.

By invoking the Born-Oppenheimer approximation that the solvent and intramolecular vibration are slow on the very fast time scale of the H^+ motion (a point of view implicit in the above discussion), k can be rewritten in terms of the fluctuations in the coupling C and the splitting $\Delta\varepsilon$:

$$k = (2/\hbar^2) \int_0^\infty dt \; \mathrm{Re} < C(0) e^{\frac{i}{\hbar}\int_0^t d\tau \Delta\varepsilon(\tau)} C(t) >. \qquad (6)$$

We pause to note that in the gas phase case, the proton would (often) tunnel coherently and no rate process or rate constant would exist. In this case, it is the spectroscopic tunnel splitting that is of interest and not the rate (8). By contrast, in solution the fluctuations in the coupling C and especially the splitting $\Delta\varepsilon$ due to the Q and S motions render the tunneling incoherent. Then k is well-defined.

BH have applied the formulation above to a model symmetric intramolecular reaction $AH^+ \cdots A \rightarrow A \cdots H^+A$ in a model water solvent. The model reaction system is very roughly similar to the oxonium ion $[HOH \cdots OH]^-$. The H^+ double well potential is constructed from Lippincott-Schroeder potentials (19), which describe hydrogen-bonded systems reasonably well. The coupling C is determined by a semiclassical method and involves proton barrier penetration integrals. The solvent coordinate and its electrical coupling to the H^+ transfer system are defined using a recent solution phase reaction path Hamiltonian Theory (20). The solvent is described for simplicity as a generalized dielectric continuum [see e.g. ref. (21)], some features of which are mentioned below. In the following we highlight the results and interpretation, and suppress most of the details.

The splitting is found to be mainly determined by the solvent, i.e. $\Delta\varepsilon = \Delta\varepsilon(S)$, while the coupling is found to be governed mainly by the Q vibrations and to increase exponentially $C(Q) = C_{eq} \exp[-\alpha Q]$ as Q becomes negative, i.e., as the A-A separation diminishes past the equilibrium value corresponding to $Q = 0$.

Approximate analytic expressions for the rate constant can be found from Eq. (6). For "high" temperatures such that $\hbar\omega/k_BT \lesssim 2$, BH find that

$$k = \frac{\pi^{1/2} <C^2>}{\hbar k_B T} \left(\frac{k_B T}{\Delta G^{\neq}}\right)^{1/2} e^{-\Delta G^{\neq}/k_B T} \qquad (7)$$

is accurate. In many hydrogen bonded systems, the Q vibration frequency is sufficiently low (500 cm^{-1}), so that "high" temperatures can correspond to room temperatures. The thermally averaged square coupling $<C^2>$ increases the rate as the Q fluctuations increase. On the other hand, the rate decreases exponentially with activation free energy $\Delta G^{\neq} = \Delta G^{\neq}_s + \Delta G^{\neq}_\alpha$. The solvent contribution $\Delta G^{\neq}_s$ measures the cost for the solvent to reach the point $S^{\neq}$ in Fig. (5) where tunneling occurs, while $\Delta G^{\neq}_\alpha$ is related to the cost of Q fluctuations to reach small AA separations where the coupling is larger. (A different analytic form is found in BH theory at "low" temperatures $\hbar\omega/k_BT \gtrsim 2$.)

The temperature dependence of this result for k is of the form $k \sim \exp[BT - E_a/k_BT]$, with the exp (BT) contribution arising from the increase of the coupling fluctuations $<C^2>$ with increasing T as the Q vibrations increase in amplitude. The activation energy $E_a = \Delta G^{\neq}_s + \Delta G^{\neq}_\alpha$, it should be noted, is <u>not</u> given by the barrier height of the potential in the proton coordinate q; instead it is determined by the solvent and intramolecular vibration fluctuations. This means that measured activation energies cannot be used to identify proton barrier heights.

On the other hand, it is the proton barrier in q through which the proton tunnels. The kinetic isotope effect (KIE), k_{H^+}/k_{D^+}, for H^+ and D^+ transfer is dominated in Eq. (7) by the different mean square coupling fluctuations $<C^2>$ for the two isotopes. This factor can be expressed in terms of the value C_{eq} of the coupling at the equilibrium separation Q=0 and the fluctuations,

$$<C^2> = C_{eq}^2 e^{2\alpha^2<Q^2>} = C_{eq}^2 \exp\left(\frac{2\alpha^2 k_B T}{m\omega^2}\right), \quad (8)$$

where m is the reduced mass of the intramolecular Q vibration. C_{eq}^{H+} for a proton exceeds C_{eq}^{D+} for a deuteron, since the lighter proton can tunnel more readily through a given double well potential barrier in q. On the other hand, the thermal fluctuations in Q favor D^+ tunneling over that for H^+. The reason for this is the fact that α_{D+} exceeds α_{H+} since the coupling C(Q) and thus the tunneling probability rises more quickly as Q contracts for D^+ than for H^+. This is a reflection of the feature that the wavefunction for the more massive D^+ varies more rapidly in space than for the lighter H^+. Since the coupling fluctuations grow as the temperature increases, the KIE declines as T rises. It is to be noted that this description of isotopic effects differs markedly from the standard conception related to differing zero point energies for the reactants and the transition state.

Table III. Kinetic Isotope Effects

R_{eq}(Å)	Δq(Å)	$\kappa_{H^+}/\kappa_{D^+}$ [a)]	$\kappa_{H^+}/\kappa_{D^+}$ [b)]
2.7	0.7	7.7	170
2.8	0.8	30	530
2.9	0.93	190	2200
3.0	1.1	1530	10^4

a)Hydrogen bond properties included.
b)Hydrogen bond properties ignored.

Calculations using the BH theory show that the KIE k_{H+}/k_{D+} is very sensitive to the hydrogen bonding potential. To emphasize this, Table III shows the results of a model calculation which is based on a Lippincott-Schroeder H-bonding potential and which accounts for the key H-bond features of rapid drop of both the central barrier height in q for the proton and the proton frequency in the reactant well. As the equilibrium separation X_{eq} between the flanking A groups is decreased and the proton is transferred over correspondingly smaller distances Δq, the KIE declines precipitously. This is due to the presence of the hydrogen bond which results in greater coupling C as X_{eq} diminishes. By contrast, Table III also shows the KIE's calculated when these special features of H bonding (dropping barrier height and proton frequency) are simply ignored, as they are in some treatments (15). This neglect results in very large overestimates of the KIE. These features emphasize the critical importance of having reasonable quantum chemical input in any study of proton transfers.

Finally, we note that proton transfer reactions share some features with electron transfers (21, 22); in particular, there is a dependence on solvent fluctuations which modulate the splitting. The key new feature that is crucial for proton transfers is the coupling variation. This is much more important for H^+ transfers than for say outer sphere electron transfers due to the greater mass of H^+ compared to the electronic mass.

Acknowledgments

We thank D. Borgis, S. Lee, M. Ferrario, G. Ciccotti, R. Kapral, J. Bergsma, B. Gertner and K. Wilson for their collaboration in some of the work reported here. Acknowledgment is made to the donors of the Petroleum Research Fund, as administered by the American Chemical Society. This work was also supported in part by NSF grants CHE84-19830 and CHE88-0852.

References

1. See e.g. C. Reichardt, Solvent Effects in Organic Chemistry (Verlag Chemie, Weinheim, 1979).
2. G. van der Zwan and J. T. Hynes, J. Chem. Phys. **76**, 2993 (1982); ibid., **78**, 4174 (1983); Chem. Phys. **90**, 21 (1984).
3. R. F. Grote and J. T. Hynes, J. Chem. Phys. **73**, 2715 (1980).
4. J. T. Hynes, in The Theory of Chemical Reaction Dynamics, edited by M. Baer (CRC Press, Boca Raton, FL, 1985), Vol. IV, p 171.
5. J. Bergsma, B. J. Gertner, K. R. Wilson and J. T. Hynes, J. Chem. Phys. **86**, 1356 (1987); B. J. Gertner, J. P. Bergsma, K. R. Wilson, S. Lee and J. T. Hynes, ibid., **86**, 1377 (1987).
6. B. J. Gertner, K. R. Wilson and J. T. Hynes, J. Chem. Phys. (in press).
7. J. Chandrasekhar, S. F. Smith and W. L. Jorgensen, J. Am. Chem. Soc., **106**, 3049 (1984); **107**, 154 (1985).
8. J. C. Keck, Discuss. Far. Soc. **33**, 173 (1962); J. B. Anderson, J. Chem. Phys. **38**, 4684 (1973); C. H. Bennett, in Algorithms for Chemical Computations: ACS Symp. Ser. **46**, ed. A. S. Nowick and J. J. Barton (American Chemical Society, Washington, DC, 1977). p 63.
9. S. H. Northrup and J. T. Hynes, J. Chem. Phys. **73**, 2700 (1980).
10. D. Chandler, J. Chem. Phys. **68**, 2959 (1978); R. O. Rosenberg, B. J. Berne and D. Chandler, Chem. Phys. Lett. **75**, 162 (1980).
11. H. A. Kramers, Physica (The Hague). **7**, 284 (1940).
12. G. Ciccotti, M. Ferrario, J. T. Hynes and R. Kapral, to be submitted.
13. See, e.g., S. Scheiner, Acc. Chem. Res. **18**, 174 (1985); B. O. Roos, Theor. Chim. Acta. 42, 77 (1976). P. Schuster, in The Hydrogen Bond ed. by P. Schuster, G. Zundel and C. Sandorfy (North-Holland, Amsterdam, 1976). Vol. 1, p 25.
14. D. Borgis and J. T. Hynes, to be submitted.
15. R. R. Dogonadze, A. M. Kuznetzov and V. G. Levich, Electrochim. Acta **134**, 1025 (1968); E. D. German, A. M. Kuznetzov and R. R. Dogonadze, J. Chem. Soc. Faraday 11, **76**, 1128 (1980).
16. See e.g. the paper by S. Scheiner in Ref. 13.

17. S. Lee, D. Ali and J. T. Hynes, to be submitted.
18. See e.g. T. Carrington and W. H. Miller, J. Chem. Phys. **84**, 4364 (1986).
19. E. R. Lippincott and R. Schroeder, J. Chem. Phys. **56**, 1099 (1955).
20. S. Lee and J. T. Hynes, J. Chem. Phys. **88**, 6853, 6863 (1988).
21. J. T. Hynes, J. Phys. Chem. **90**, 3701 (1986).
22. See, e.g., L. D. Zusman, Chem. Phys. **49**, 295 (1980); D. F. Calef and P. G. Wolynes, J. Phys. Chem. **87**, 3387 (1983); H. Sumi and R. A. Marcus, J. Chem. Phys. **84**, 4894 (1986); I. Rips and J. Jortner, J. Chem. Phys. **87**, 2090 (1987).

CONCEPTS IN SURFACE SCIENCE AND HETEROGENEOUS CATALYSIS

Roger M. Nix and Gabor A. Somorjai
Center for Advanced Materials, Lawrence Berkeley Laboratory and Dept. of Chemistry, University of California, Berkeley, California 94720, U.S.A.

Abstract

The development and application of surface sensitive techniques over the last two decades has greatly improved our understanding of surface phenomena at the molecular level. From this newly acquired vast data base new concepts have emerged and it is the purpose of this paper to review those that relate most directly to the chemical properties of surfaces. Concepts concerned with the structure of clean and adsorbate-covered surfaces, the nature of the surface chemical bond, the dynamics of gas–surface interactions, surface chemical reactions and catalytic processes are presented and illustrated with examples from the recent literature.

1. Introduction

The study of surface phenomena is a multi-disciplinary science, utilizing a wide spectrum of experimental and theoretical techniques. The field encompasses research not only on the chemical properties of interfaces but also the mechanical, optical, electrical and magnetic properties. During the past twenty years there has been an explosive growth in the study of surfaces and interfaces because it is one of the intellectual frontiers of the physical sciences and is also of immense technological importance. From the results of these molecular level studies, new concepts have emerged which have markedly altered many of our views of surface phenomena. The purpose of this paper is to review some of the more general concepts that have emerged from this research in the field of surface science and to present examples that both illustrate the phenomena and demonstrate their role in more complex systems. We shall, however, concentrate on those particular features of surface structure and molecular behaviour at interfaces which are most directly related to the chemical properties of the surface.

One field which has benefitted greatly from the advances in our fundamental understanding of chemical phenomena at surfaces is that of catalysis science. Indeed, as a subfield of surface science, the study of heterogeneous catalysis has undoubtedly been of unique importance. For this reason we have, in many cases, chosen examples that show how these concepts relate to catalytic phenomena and catalyst systems that are of major importance in the chemical and petroleum industries. In addition, we include a section which summarizes concepts that have emerged from studies directed specifically at the catalytic properties of surfaces.

J. Jortner and B. Pullman (eds.), Perspectives in Quantum Chemistry, 97–121.

2. The Experimental Approaches and Techniques of Modern Surface Science

There are two main approaches to the study of surfaces. The first utilizes high surface area materials to enhance the contribution of the surface to the overall properties and characterizable features of the solid. One class of solids that fall into this category are the microporous crystalline zeolites[1,2] and related materials which may have surface areas in the range 200-700 m^2/g: an example of a zeolite structure is shown in Fig.1. This particular zeolite is a naturally occurring mineral but, since the discovery that these materials possess unique properties in the role of both catalysts and selective adsorbents (molecular sieves) several hundred such structures have been synthesized in the laboratory; many of which have no analogue in nature. Indeed, the progress in this field has been such that novel structures tailored to the needs of particular chemical processes have been prepared . The zeolite illustrated in Fig.1 contains only silicon, aluminum and oxygen framework atoms but in recent years compositions including many other elements (eg Ga, Ge, Fe, P) have been prepared and new classes of compounds, such as the aluminum phosphates[3], have also been synthesized in microporous, crystalline forms. Microporous, crystalline solids may be studied by electron microscopy , solid state NMR, EXAFS, X-ray and neutron diffraction in order to determine the location, coordination number and chemical environment of each atom - many of which, by the very nature of the materials, are surface atoms. This information can then be used, for example, to relate molecular structure to catalytic behaviour.

The second approach involves the use of model low surface area specimens. These consist of single crystal or polycrystalline samples with a surface area typically of the order of

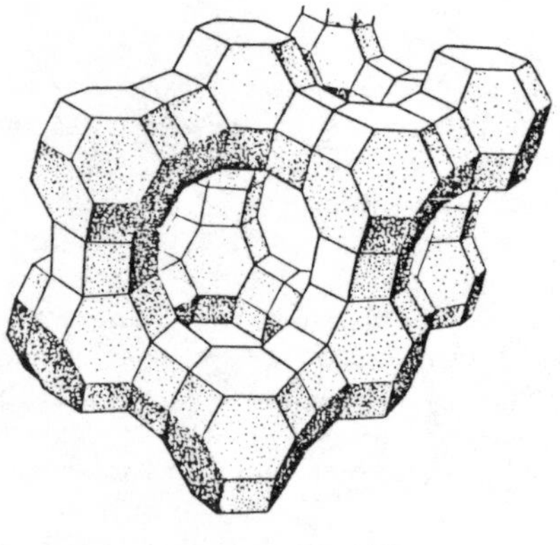

Figure 1: Structure of Faujasite - a naturally occurring Zeolite.

Table I

Techniques of Modern Surface Science

- **Electron–Surface Scattering**

Electron Spectroscopy	- Auger Electron Spectroscopy (AES)
	- Ultraviolet Photoelectron Spectroscopy (UPS)
	- X-ray Photoelectron Spectroscopy (XPS)
	- Inverse Photoemission Spectroscopy (BIS)
	- Electron Energy Loss Spectroscopy (HREELS)
Electron Diffraction	- Low Energy Electron Diffraction (LEED)
Electron Microscopy	- Scanning Auger Microscopy (SAM)
	- Scanning Electron Microscopy (SEM,STEM)
	- Transmission Electron Microscopy (TEM,STEM)
	- Reflection Electron Microscopy (REM)
Tunnelling Microscopy	- Scanning Tunnelling Microscopy (STM)

- **Photon–Surface Scattering**

Spectroscopy	- Infra-Red Spectroscopy (IR,FTIR)
	- Raman Spectroscopy
	- Nuclear Magnetic Resonance (NMR)
	- X-Ray Absorption (EXAFS,SEXAFS,XANES)
	- Laser Techniques (SHG,SFG)
X-Ray Diffraction	- Grazing Angle X-Ray Diffraction

- **Molecule/Ion–Surface Scattering**

Molecular Beam Scattering	- Reactive Molecular Beam Scattering (RMBS)
	- Thermal Helium Scattering
Ion Scattering	- Secondary Ion Mass Spectrometry (SIMS)
	- Ion Scattering Spectroscopy (ISS)

- **Other Techniques**

Chemisorption Techniques	- Temperature Programmed Desorption (TPD)
	- Temperature Programmed Reaction Spectroscopy (TPRS)

Work Function Measurements
Radiotracer and Isotopic Labelling

$1cm^2$. Well defined surfaces can be prepared by the careful orientation, cutting and polishing of single crystals which may then be cleaned by ion sputtering and other techniques in a high vacuum system. It is much easier to control and determine the cleanliness, structure and composition of these $1cm^2$ samples than high surface area materials and thereby more definitive measurements of relevant atomic and molecular level parameters may be obtained.

A wide range of techniques have been developed that are capable of specifically probing the properties of interfaces[4]. A selection of these experimental techniques is presented in Table I - they predominantly involve the use of photons, ions and low energy electrons to probe the immediate and near surface regions. The combined use of several of these techniques provides complimentary information on different aspects of the interface including composition (AES, XPS, ISS), geometric structure (LEED, XRD, ISS, TEM[HREM], XPD, STM), electronic structure (UPS, XPS, BIS) and adsorbate bonding (HREELS, LEED, TPD, XANES). Some of these techniques can also be used to look at the surfaces of high surface area solids but, in many cases, readily interpretable information can only be extracted from single crystal studies. Furthermore, most of these probes can only be used in a high vacuum environment.

In the study of catalytic surface phenomena low surface area specimens have proved to be a very valuable tool, especially as models for supported metal catalysts. This very important class of catalysts consist of small (10-1000Å) metal particles dispersed on a support; typically, but not exclusively, an oxide. The metallic surface areas of such materials generally fall in the range 1-100 m^2/g. The study of such systems using surface sensitive techniques is restricted in part, as mentioned above, by the inherent limitations of many of the experimental probes but more fundamentally by the complex, heterogeneic nature of the materials themselves. Judiciously chosen, single crystal model samples allow fundamental studies of reaction mechanisms and the effects of surface structure and bonding to be performed under well defined conditions. It is undoubtedly in this area of catalysis that modern surface science has contributed the most.

The challenge to the surface scientist working in applied technological fields such as catalysis is to relate the properties of these low surface area samples to those of real systems under their operating conditions. The problem revolves around the "pressure gap" - the application of most surface sensitive techniques is restricted to high vacuum conditions and typically involves measurements at relatively low temperatures and coverages. In contrast, the process that is being modelled often occurs under conditions of high pressure, temperature and coverage. One approach to this problem involves the use of UHV apparatus equipped with environmental cells in which conditions much closer to those actually employed can be attained. This has been successfully applied to the study of both electrochemical and catalytic phenomena. In the latter case, the low area single crystal specimen may, after preparation and characterization under high vacuum conditions, be enclosed in an isolation cell and then exposed to reactant gases at elevated pressures[5-8](Fig.2). The rate and kinetic parameters of the reaction, along with the selectivity, can be ascertained from the product distribution which is in turn determined using mass spectrometry or gas chromatography. After reaction the sample is transferred back into high vacuum and the surface composition and structure redetermined. Therein lies a method for correlating high pressure catalytic behaviour with specific surface properties. A few examples of this powerful technique will be presented later, but first we will review some of the more classical concepts of surface chemistry as well as concepts that have emerged from fundamental studies in modern surface science.

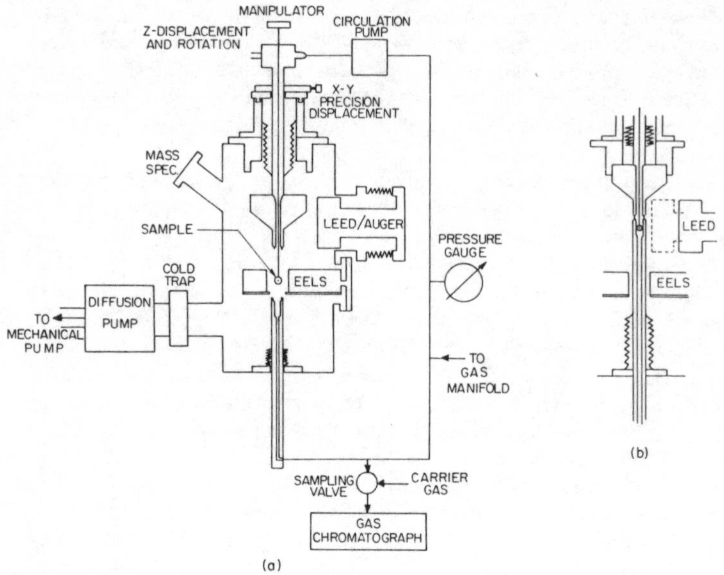

Figure 2: Schematic Illustration of the Design of a Combined UHV/High Pressure Apparatus.

3. Classical Concepts in Surface Science

Three of the classical concepts in surface chemistry are:

- the surface free energy and surface segregation.
- the surface space charge.
- the unique properties of curved surfaces.

The fact that **the surface free energy** is always positive means that any condensed phase will preferentially minimize its surface area and/or place on the surface that material which has the lowest free energy. As a general guideline[9], metals have higher surface free energies than oxides, so oxides usually cover the metals. Since water has an even lower surface energy than oxides, it covers the oxides, while organic molecules such as benzene have a lower surface energy still and will cover the aqueous phase. Finally, fluorocarbons have the lowest surface energy of all and are therefore the most specific to surfaces. Those materials that segregate to surfaces because of their low surface energy are called surface active agents. Their behaviour, for example, underlies the principle of operation of detergents.

The theoretical foundations for the thermodynamic description of **surface segregation** were developed by Gibbs[10]. In dilute alloy systems, for example, segregation is driven predominantly by the lowering of the surface free energy and relief of bulk strain energy[11–13], although other factors (eg heat of mixing of alloy phases) can modify the extent of segregation. The important conclusion, however, is that the driving force to minimize the total free energy of the system may give rise to surface compositions very different from the bulk even in what are nominally single phase systems and, in doing so, may impart to the surface superior mechanical or chemical properties. Modern surface science studies on alloy systems have not only confirmed the basic premise of surface segregation on an atomic scale but have also allowed a detailed and quantitative test of the various theoretical models (eg. ref.14).

The next figure (Fig.3) shows how a **space charge** may buildup at any interface due to the extension of the electronic wavefunctions out of the solid into the vacuum or other medium - the electrons 'spill out' leading to a surface dipole and a space charge whose thickness decreases with increasing initial charge density[15]. The resulting charge sepa-

$$(P_{ext} - P_{in}) = \frac{2\gamma}{r}$$

$$\ln \frac{P_r}{P_{r=\infty}} = \frac{2V_m\gamma}{rRT}$$

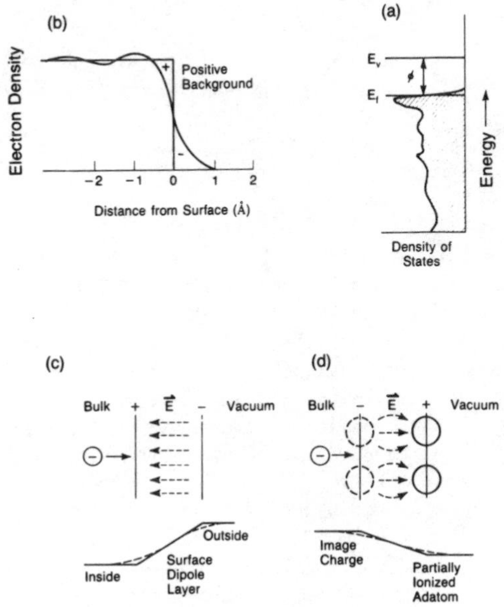

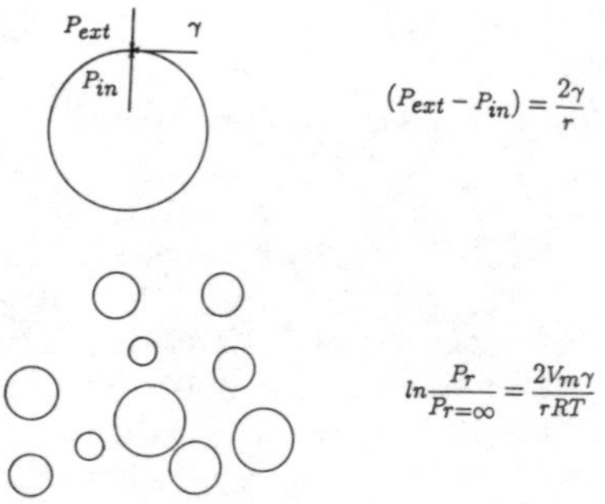

Figure 4: Thermodynamic Properties of Small Spherical Particles
γ - surface tension,
r - particle radius,
V_m - molar volume.

Figure 3: The Electronic State of the Surface: (a) The Fermi Energy (E_F) and Work Function (ϕ), (b) Electron Density Distribution and the Formation of a Surface Dipole, (c,d) The Dipolar Contribution to the local Work Function.

ration provides a static electric field that is an important property of any surface and a face specific contribution to the work function. Furthermore, the adsorption of atoms or molecules that donate or accept electrons can drastically modify this surface dipole. Adsorbate bonding and diffusion, as well as surface ionization are all influenced by this surface electric field.

Curved surfaces have vapor pressures that are different internally and externally in order to maintain the surface curvature (Fig.4A). Particles with different radii of curvature also have different vapor pressures and solubilities[16,17]; more specifically, the smaller the particle, the higher its vapor pressure or solubility Fig.4B). As a result, in a system containing a mixture of small and large particles the small particles will dissolve preferentially. Similarly, large particles will grow at the expense of small ones (Ostwald ripening) thus leading to phenomena such as sintering. Nature has a way of preventing this by providing systems where all the particles have very similar radii. Examples of these include colloidal systems (eg milk and blood) in which the particles can also stabilized against coalescence by electrostatic repulsion between electrical double layers induced by the surface charge separation described above.

We can therefore expect the behavior of surface systems to be markedly dependent on their surface energies, surface charge and size; indeed these concepts are used in areas such as catalysis science (to prepare, stabilize and regenerate catalyst particles), the paint industry and separation technology.

4. Modern Concepts in Surface Science

Modern surface science has provided many new concepts, some intimately related to those already mentioned, and brought our understanding of the properties of surfaces down to the molecular and even atomic levels.

4.1 Relaxation, Reconstruction and Atomic Scale Structure (Terraces, Steps and Kinks) of Clean Surfaces

Since the 3D periodicity of a solid is terminated at the surface, it is possible for the interlayer spacing of atomic layers near the surface to differ from that found throughout the bulk ("**relaxation**"). Surface crystallography studies[18] have shown that in vacuum virtually all clean metal surfaces relax and that the spacing between the first and second atomic layers is significantly (ca.1-20 %) reduced from that which characterizes the bulk. The lower the atomic packing and density of the surface, the larger is the relaxation.

The forces which lead to relaxation of surfaces and result in a change in the equilibrium position and bonding of surface atoms can give rise to more drastic **reconstruction** of the outermost layers; that is, the surface can assume an atomic structure which differs more fundamentally from that expected from termination of the bulk structure. One example is shown in Fig.5. The gold, platinum and iridium (100) surfaces all show reconstruction: the surface unit cell which would be square in the absence of reconstruction is, instead, pseudo-hexagonal[19]. The surface structure assumed involves not only closer packing but also buckling of the surface layer. Many other surfaces of monatomic solids also exhibit reconstruction and this can lead to unique electronic and chemical properties. The reconstruction of semiconductor interfaces is generally more dramatic than is the case for metals. An example is shown in Fig.6 which illustrates the (2x1) surface reconstruction of the Si(100) face. Extensive analysis of LEED and ion scattering data[20,21] has led to the

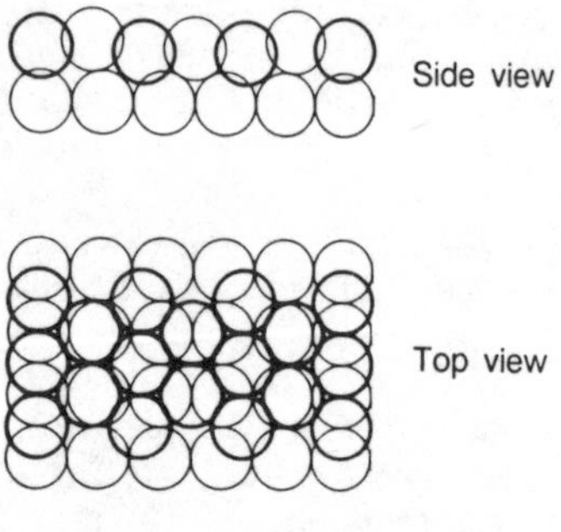

Figure 5: Ir(100) - (5x1).

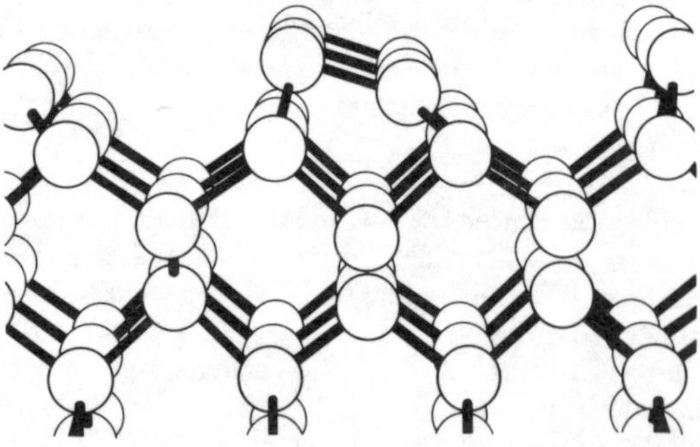

Figure 6: Si(100) - p(2x1).

structure shown, in which the outermost atomic plane consists of buckled but untwisted dimers, and relaxation extends down to the fourth or fifth layer.

The phenomena of relaxation and reconstruction are both microscopic expressions of the minimization of the surface free energy of the system.

The presence of **atomic steps and kinks**, even on nominally perfect low index crystal faces, has been revealed by several imaging techniques (eg low energy electron microscopy[22], photoelectron microscopy[23], reflection and transmission electron microscopy[24]) but recent developments in Scanning Tunnelling Microscopy (STM) in particular have greatly increased our atomic-level understanding of local surface structure[25,26]. Fig.7 is an STM image of a rhenium (0001) surface[27] that was passivated by adsorption of half a monolayer of sulphur, thereby making it resistant to oxidation or other chemical attack (by this means it could be studied by a scanning tunnelling microscope even in air). From this picture, the presence of kinks and steps can clearly be seen. Moreover, it has become clear that the density of step atoms on even the lowest energy surfaces can be relatively high and that such features will be stable under virtually all experimental conditions including those pertaining to heterogeneous catalysis.

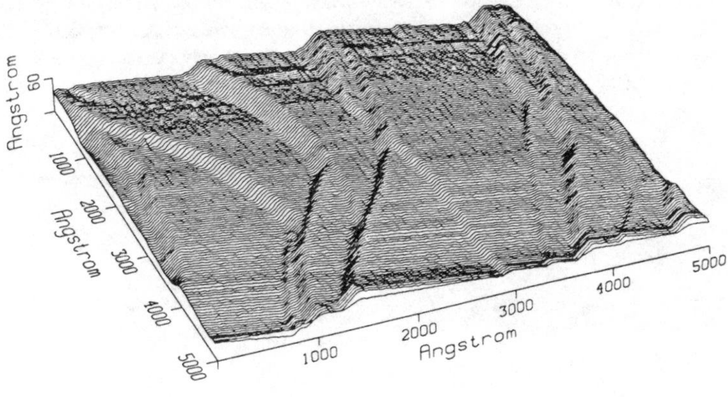

Figure 7: 3-D Projection of STM Data for a Sulfur-passivated Re (0001) Basal Plane showing Terraces separated by Steps of 1-10ML (vertical scale x5).

The electronic properties of the step atoms differ markedly from those of the terrace atoms. This is reflected in the decrease in the average work function with increasing step density on vicinal surfaces[28] but is more directly illustrated by measurements of the local barrier height using a STM[27]. Intimately related to the change in barrier height are variations in the electric field strength at steps. It is not surprising therefore that such sites are implicated in many aspects of adsorption, desorption, bond dissociation etc.[29–32].

The unusual properties of curved solid surfaces (small particles) are clearly also related to the presence of such atomic-scale structure. Recent work on small metal clusters has shown,however, that in addition to the unusual coordination of surface atoms in this size regime, such particles may also exhibit fluxionality of structure[33], markedly different electronic properties (eg. work function and electronic affinity[34]) and substantially different phase transition temperatures (eg. melting points[35]).

4.2 Adsorption-Induced Changes in Structure and Composition

A related concept that has emerged from modern surface science studies is that of adsorbate-induced restructuring and segregation. In order to demonstrate these phenomena let us consider several examples. When strong bonds are formed between an adsorbate and a surface, the surface atoms may modify their positions to conform to the new chemical environments - this is the phenomenon of **adsorbate-induced surface reconstruction.** For example, low coverages of hydrogen on W(001) induce a c(2x2) surface reconstruction at 300K[36,37]. Similarly, the presence of a quarter monolayer of atomic carbon on Ni(100) induces a reconstruction of the topmost nickel atoms both parallel and perpendicular to the surface, in such a manner that the four nickel atoms surrounding each carbon atom are rotated with regard to the underlying layers[38] (Fig.8). The importance of adsorbate induced restructuring such as this in heterogeneous catalysis should not be underestimated since the catalytically active surface may only exist in the presence of certain adsorbates. The time scale for the reconstruction may be much shorter than catalytic turnover times or, in other cases, may even be determined by the reaction mechanism. There have been numerous investigations of kinetic oscillations in the catalytic oxidation of CO on Pt surfaces[39]. These self-sustained reaction rate oscillations may be accompanied by large temperature changes as shown in Fig.9 and have been observed over a wide range of conditions. One of the mechanisms shown to operate under low pressure, isothermal conditions[40] involves the restructuring of the Pt(100) surface in the manner described in the previous section. In the

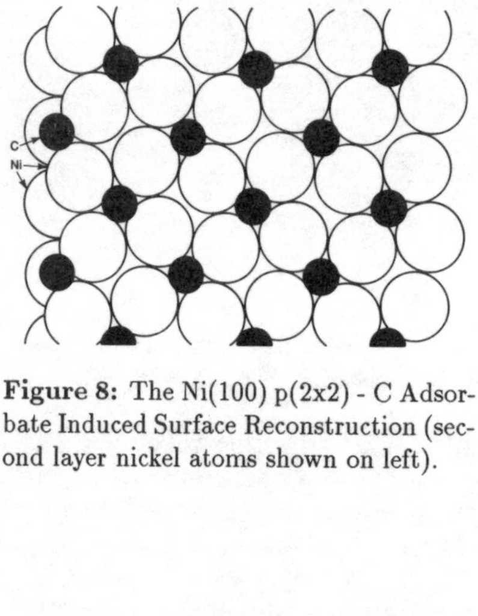

Figure 8: The Ni(100) p(2x2) - C Adsorbate Induced Surface Reconstruction (second layer nickel atoms shown on left).

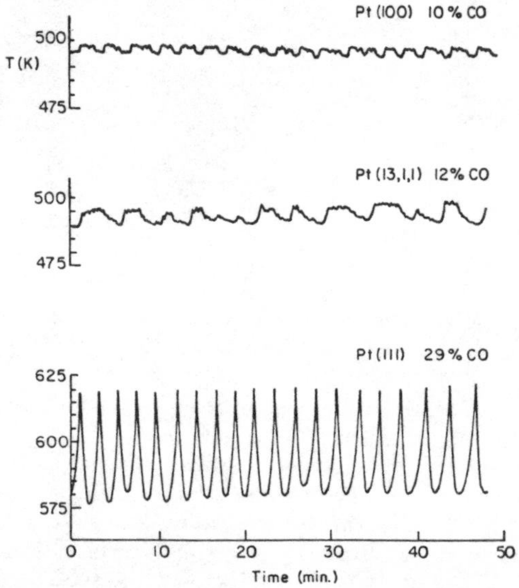

Figure 9: Self Sustaining Temperature Oscillations in the CO Oxidation Reaction over different Platinum Single Crystal Surfaces (from ref.41).

presence of a high concentration of adsorbed carbon monoxide the primitive (1x1) surface structure with a square unit cell is preferred, whereas in the presence of atomic oxygen the reconstructed hexagonal surface structure is more stable. Variations in the surface concentration of the adsorbed species during the oxidation, oscillations in the reaction rate and restructuring of the platinum surface are all intimately coupled and occur on the same time scale thereby yielding the observed behaviour. This, however, is only one of several mechanisms that can lead to oscillatory behaviour and the nature of the driving force at higher pressures is still under debate[41–43].

If reconstruction occurs very slowly, on a time scale that is much longer than that of the reaction, there may be long term changes in the catalytic reaction rates (ie either a gradual increase in activity or a slow poisoning of the catalytic reaction). One such example of a slow diffusion-controlled reconstruction is the coalescence of atomic surface steps into multi-atomic steps that ultimately leads to facetting[44]. This type of behaviour is likely to be irreversible under reaction conditions but might be reversed by thermal restructuring upon desorption of the reaction intermediates. A related phenomenon is the change in shape, structure and size of small supported metal particles upon oxidation or reduction. This effect is utilized in the regeneration of many aged catalysts where an oxidation/reduction cycle can result in an enhanced dispersion of the active phase[45–47]. Oxidation of the large metal particles formed by sintering generally leads to better wetting of the underlying oxide surface. During subsequent low temperature reduction there is a tendency for the oxidized layer to break up into smaller metal particles: hence the increase in metal surface area and dispersion. In certain systems this procedure may be repeated many times, reproducibly yielding the same catalyst particle structure and therefore the same initial catalytic activity.

The examples above all concern cooperative reconstruction of the substrate surface on a microscopic scale (~ 50–10000 Å): it can also occur, however, on a very local scale. For example, recent diffuse LEED work[48] has shown that the presence of an adsorbed oxygen atom in a four-fold hollow site on a W(100) surface induces a clustering (or local reconstruction) of the neighbouring tungsten atoms.

It was described earlier how alloy systems may minimize their surface free energy by segregation of one of the components to the surface. The surface composition that satisfies the criterion in the presence of reactive gases will, however, frequently differ from that which holds at a vacuum interface. In certain cases, segregation will result from the preferential reaction of one component to yield a phase insoluble in the alloy. This is well documented for alloys in which selective oxidation of one component can occur; particular examples include alloys of Pt with other Gp.VIII metals[49,50] and alloys of electropositive elements (eg. Zr, Ti, lanthanides and actinides) with more noble metals[51]. A more subtle and surface specific demonstration of this concept is where segregation is induced purely by the different heats of adsorption of a gas phase constituent on the two components of the alloy. This phenomenon is not well documented in the literature although, for example, Bouwman et al.[52] report a reversible surface segregation of Pd in polycrystalline Ag-Pd alloys upon CO chemisorption and a theoretical treatment of such **chemisorption induced segregation**, incorporating some comparisons with experiment, has been carried out by Tomanek et al.[53].

A closely related phenomenon appears to be responsible for at least some aspects of the SMSI (strong metal-support interaction) exhibited by certain metal-oxide systems. In particular, reduction of titania supported catalysts at high temperature can lead to migration of a partially reduced titania species onto the metal particles, thereby blocking the low temperature adsorption of CO and H_2, but not preventing the hydrogenation of CO at higher temperatures[54–56]. The effect may be reversed by annealing in an oxidizing environment.

4.3 Epitaxial Growth and Surface Compound Formation

Fundamental studies of the growth of evaporated films on a multitude of different substrates have led to the concept of **epitaxial growth.** In its broadest interpretation this concept covers all cases where the substrate acts as a 'template' and has a significant influence on the growth mode of the deposited material. A much more restricted definition would include only those examples where the substrate actually imposes its own crystal structure, orientation and lattice parameter on the adsorbed overlayer (ie 'pseudomorphic' growth). The idea is best illustrated by reference to the many studies of ultrathin metal overlayers on metal single crystals[57,58]. There are numerous instances where the growth mode of one metal on another varies according to the orientation and symmetry of the exposed substrate crystal face: a far better illustration of this concept, however, may be obtained from examples such as the Pd/Ag(100) and Fe/Cu(100) systems. In the former case[59], where the two metals have the same bulk structure (fcc), the Pd initially grows in perfect epitaxy with a 5.1% lateral expansion of the interatomic spacing imposed by the substrate. This strained layer-by-layer growth persists to beyond 3 monolayers before relaxation to the bulk structure is seen. In the latter case the effects of the interfacial interaction are more dramatic. The Cu(100) substrate forces the iron to adopt an epitaxial fcc structure (as opposed to the bcc structure of bulk iron) up to film thicknesses of 5ML, after which the epitaxial relationship is unable to sustain the close packed Fe and collapse to a structure more closely approaching that of the bulk occurs[60]. The effect of epitaxial relationships are also evident in more complex systems in the form of preferential crystallite orientation. For example, electron microscopy studies of Cu/ZnO methanol synthesis catalysts[61] show a strong preference for structural registry of the (211) planes of copper particles with the $(10\bar{1}0)$ plane of ZnO.

Another important concept to consider is that of **surface compound formation.** The studies of Sinfelt and coworkers have shown that when particle sizes become very small and dispersions tend to unity (that is, virtually every atom is present at the surface), alloy systems exhibit very different "phase diagrams" from those that characterize the bulk systems. For example, microclusters containing both Cu and Ru or Cu and Au atoms can be produced despite the elements being completely immiscible in three dimensions[62]. This might have been predicted since the relative importance of the surface and bulk free energy contributions to the total energy of the system changes dramatically at high dispersions. Few studies, however, have directly addressed the influence of particle size on the surface composition of small particles[63]. The application of modern surface sensitive techniques to the study of model bimetallic systems on single crystal substrates has been more extensive. It should be noted that the results of such studies may not necessarily relate to the situation at very high dispersions due to the very different surface:bulk ratios in the two cases. Nevertheless, this work has been very useful in elucidating the properties of interfacial compounds. An example of such a surface compound is shown in the next figure (Fig.10). An α-Cu/Al alloy single crystal with a bulk concentration of 16 at% Al exhibits no long range order in the bulk; the surface, by contrast, is completely ordered as shown in the figure[64]. Furthermore, due to aluminum segregation, the ordered surface phase contains equal numbers of copper and aluminum atoms. Thus the surface has both a very different structure and composition from the bulk.

There is no reason why surface compound formation should be restricted to either metal-metal or gas (vacuum)-solid interfaces. Although substantially less experimental documentation of surface specific metal-oxide or oxide-oxide compound formation is available; dissolution of oxide layers into metallic substrates[65,66], chemical interaction of metals

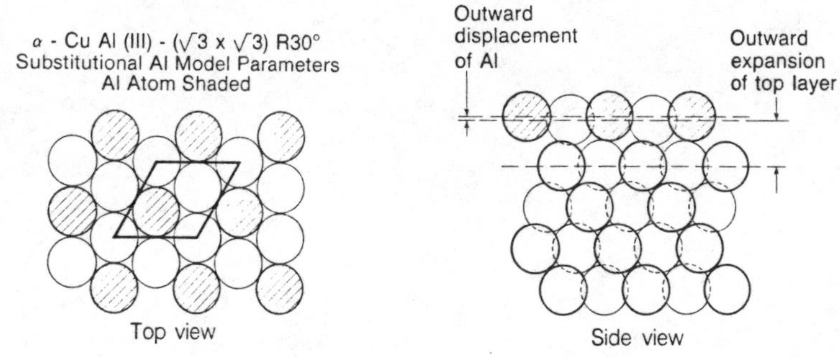

Figure 10: Surface Compound Formation on the (111) Face of an α-CuAl (16% Al) Alloy.

with oxide substrates[67,68] and mixed oxide surface compounds[69,70] have all been proposed on the basis of surface science studies of both model and complex systems.

4.4 The Surface Chemical Bond - Bonding Geometries, Thermal Activation and Coadsorption

The next concept of modern surface science to be discussed is that of the **surface chemical bond.** The binding of surface species has been found to be 'cluster-like'; this is a particularly useful concept since it permits one to use localized bonding models in the study of surfaces. It is also an approach frequently adopted in theoretical calculations of molecular adsorption[71,72]. Several organic molecules and molecular fragments that have been identified on metal surfaces by a combination of high resolution electron energy loss spectroscopy and low energy electron diffraction are shown in Figs.11 & 12. These species have the same local structure and similar chemistry to those found in multinuclear organometallic clusters for which good x-ray diffraction information is available. In fact, for virtually every surface species found so far there is a cluster equivalent that has been synthesized by organometallic chemists.

Large organic molecules frequently exhibit distortions when adsorbed on metal surfaces. Benzene and closely related aromatic hydrocarbons generally lie with theirπ-ring parallel to the surface but, as shown by LEED studies, are distorted from their equilibrium gas phase geometry due to the metal-adsorbate interaction. The stronger this interaction, the larger the distortion[73] as shown in Fig.13. Similar distortions are also found in multinuclear organometallic compounds with benzene[74], such as the ruthenium-benzene complex shown in the following figure (Fig.14), although these distortions are not as large as those seen on metal surfaces (presumably because a smaller number of metal atoms are involved in the bonding in a cluster).

In the case of aromatic heterocyclic molecules the situation regarding bonding geometry is not as clear cut[75]. Fig.15 shows one of the structural configurations of pyridine on a Rh(111) surface[76]. For pyridine there exists the possibility of bonding to the surface through the π-system, via the nitrogen alone or through both the N and C2 atoms. Thus, the molecule may assume either flat or upright structures or, as in the case illustrated, with the molecular plane oriented at an angle with respect to the surface. The actual mode of bonding adopted may be dependent upon surface coverage (ie interadsorbate interaction - see below) and temperature as well as the substrate[77–79].

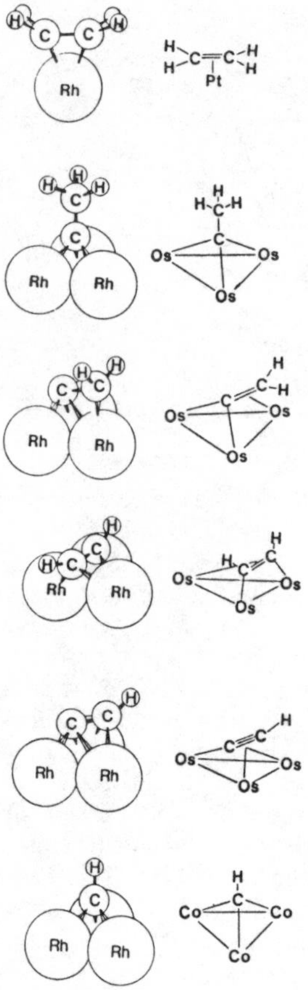

Figure 11: A Comparison of Proposed Adsorbate Species on Rh Surfaces with the binding of Isostructural Ligands in Cluster Complexes.

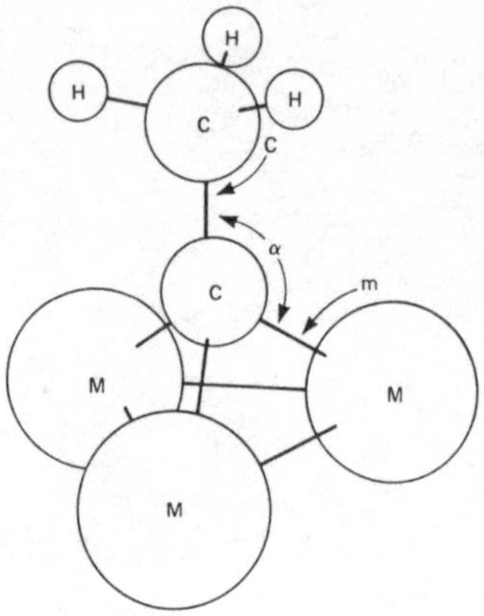

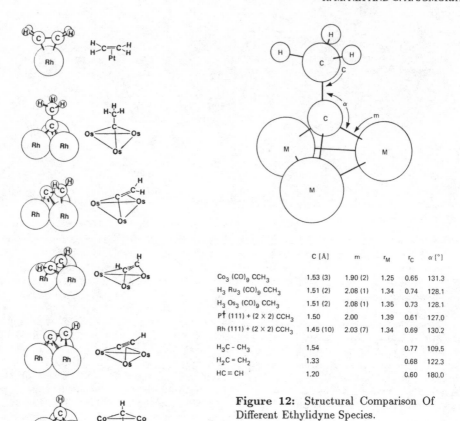

	C [Å]	m	r_M	r_C	α [°]
$Co_3(CO)_9CCH_3$	1.53 (3)	1.90 (2)	1.25	0.65	131.3
$H_3Ru_3(CO)_9CCH_3$	1.51 (2)	2.08 (1)	1.34	0.74	128.1
$H_3Os_3(CO)_9CCH_3$	1.51 (2)	2.08 (1)	1.35	0.73	128.1
Pt (111) + (2 × 2) CCH_3	1.50	2.00	1.39	0.61	127.0
Rh (111) + (2 × 2) CCH_3	1.45 (10)	2.03 (7)	1.34	0.69	130.2
H_3C-CH_3	1.54			0.77	109.5
$H_2C=CH_2$	1.33			0.68	122.3
$HC\equiv CH$	1.20			0.60	180.0

Figure 12: Structural Comparison Of Different Ethylidyne Species.

Substrate	(Gas Phase)	Pd(111)	Rh(111)		Pt(111)
Surface Structure		(3x3)-C_6H_6 + 2CO	(3x3)-C_6H_6 + 2CO	c($2\sqrt{3}$x4)rect-C_6H_6 + CO	($2\sqrt{3}$x4)rect-2C_6H_6 +4CO
The Structure of Benzene	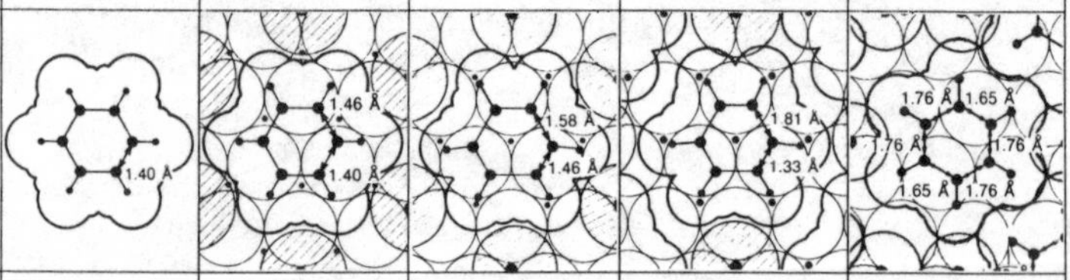				
C_6 Ring Radius (Å)	1.40	1.43±0.10	1.51±0.15	1.65±0.15	1.72±0.15
d_{M-C}(Å)	-	2.39±0.05	2.30±0.05	2.35±0.05	2.25±0.05
γ_{CH}(cm^{-1})*	670	720-770	780-810		830-850

Figure 13: The Bonding Geometry of Benzene in CO-Coadsorption Structures.

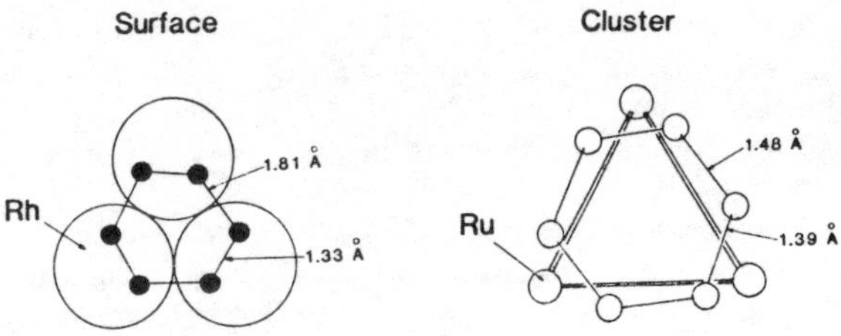

Figure 14: The Distortion of the Benzene Ring in the c($2\sqrt{3}$x4)rect-C_6H_6 + CO / Rh(111) Structure and $Ru_6C(CO)_{11}(\mu_3\text{-}C_6H_6)(\eta^6\text{-}C_6H_6)$ Complex.

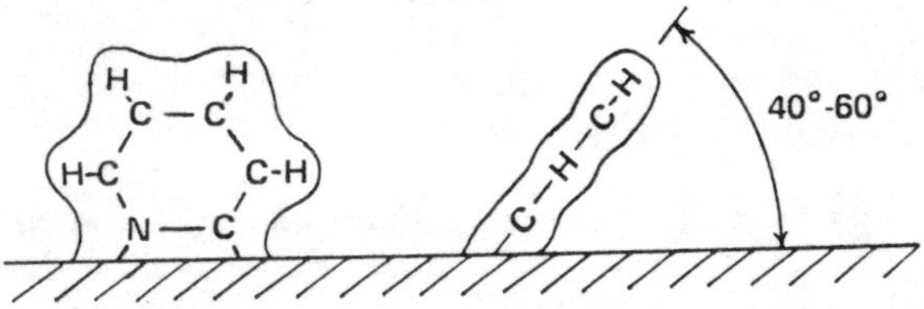

Figure 15: The Adsorption of Pyridine on Rh(111): the α-pyridyl species at 310K.

This brings us on to a closely related concept: the **thermal activation of the surface chemical bonds** (also known as temperature dependent bond rearrangement and bond activation). It is found that molecules adsorbed at low temperatures (below ca. 20K) are quite unreactive and assume geometries not unlike those in the gas phase. As the substrate is heated, unique bond breaking processes can occur within well defined temperature ranges - indeed, strong chemical bonds may be broken over very limited (~10K) temperature ranges as has been shown by temperature programmed spectroscopic studies (eg ref.80). In the case of 'complex' molecules, a progressive increase in temperature can lead to sequential bond breaking, yielding molecular fragments that are very stable within a particular temperature regime. This is demonstrated in Fig.16. Benzene and

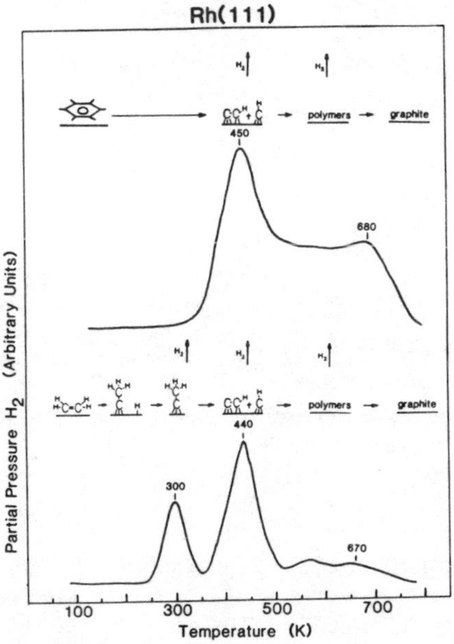

Figure 16: Thermal Decomposition Routes of Benzene and Ethylene on Rh (111).

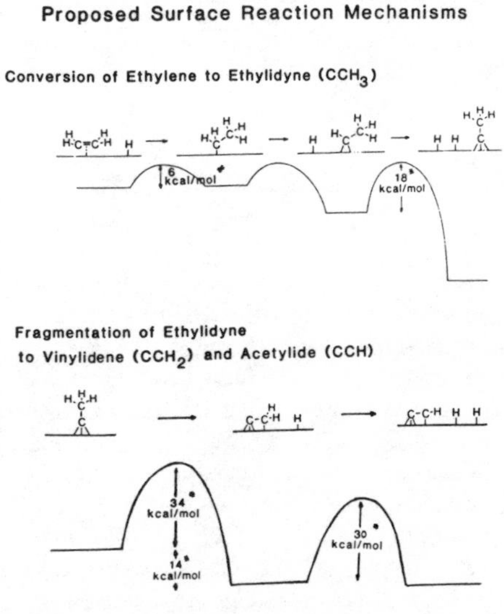

Figure 17: The Decomposition of Adsorbed Ethylene (* - from ref.87; # - from ref.88).

ethylene assume very different surface structures on Rh(111) at low temperatures: however, as the temperature is increased both molecules decompose and above ca. 450K the molecular fragments remaining on the surface are identical[76,81–83]. In fact, the adsorption of many different hydrocarbons yield surface species that are indistinguishable above a certain temperature[83,85,86].

The mechanisms of these transformations have been studied by both experiment and theory in several molecular systems. For example, studies indicate that the mechanism by which adsorbed ethylene (C_2H_4) is converted into ethylidyne (C_2H_3) involves firstly hydrogenation to a C_2H_5 intermediate with subsequent loss of two hydrogen atoms to give the ethylidyne species, rather than a direct dehydrogenation of C_2H_4 (Fig.17A). The further fragmentation of C_2H_3 to C_2H_2 and CH species can also be modelled by theory and reasonable agreement between theory and experiment exists (Fig.17B).

The next concept to come from modern surface science studies is that of the **coadsorption** bond. It is frequently found that there are large changes in the isosteric heat of adsorption with increasing coverage which lead to a marked reduction in the average heat of adsorption per molecule. This is commonly caused by a repulsive (predominantly dipolar) adsorbate-adsorbate interaction that becomes increasingly important as the interadsorbate separation decreases at higher coverages and results in a weakening of the bonding of the molecules to the surface. This is but one example of repulsion between 'like' molecules and the behaviour is well illustrated by the CO-metal systems (see eg Fig.18). In these

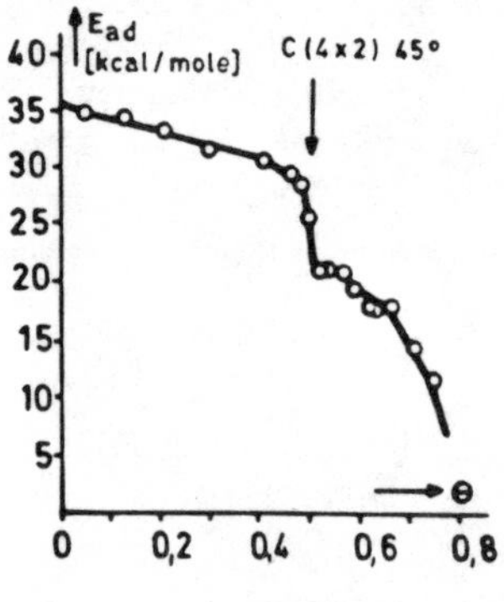

Figure 18: Heat of Adsorption for CO on the Pd(100) Crystal Face as a Function of Coverage (from ref.90).

systems there is a delicate interplay between the repulsive interadsorbate forces and structural changes within the adsorbed layer that result in modifications in the CO-substrate bonding strength and geometry. Fig.19 compares the CO/Pt(111) structure at half monolayer coverage[91], in which the CO molecules occupy well defined sites, to that observed at higher coverages on a Rh(111) substrate[92], where, to minimize mutual repulsion, the adsorbed molecules adopt a pseudo-hexagonal structure.

Clearly, because the average heat of adsorption per molecule is smaller at high coverages, the reactivity of molecules under these conditions may be very different from that at low coverages.

Attractive adsorbate-adsorbate interactions upon coadsorption of two different molecules may lead to stronger chemical bonding or pronounced structural effects. An example of the latter type is illustrated in the next figure (Fig.20). LEED and HREELS studies show that benzene molecularly adsorbs at 300K in a disordered manner on a clean Rh(111) surface[76]. It can be readily ordered, however, by coadsorption with other molecules that are electron acceptors, such as CO and NO[93–95]. Like most organic molecules, benzene

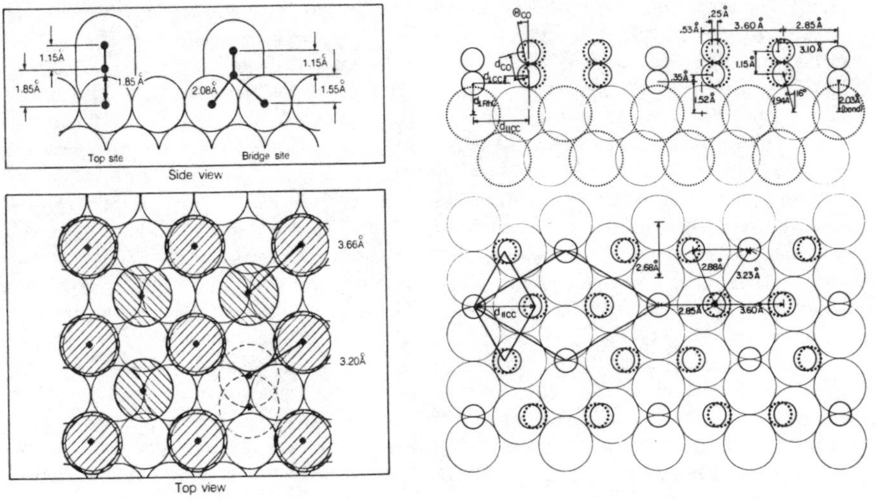

Figure 19: LEED Structures of CO-Metal Systems: A - Pt(111)-c(4x2)-2CO at T = 150K (0.65×10^{15} molecules CO/ cm^2), B - Rh(111)-(2x2)-3CO at T = 240K (1.04×10^{15} molecules CO/ cm^2).

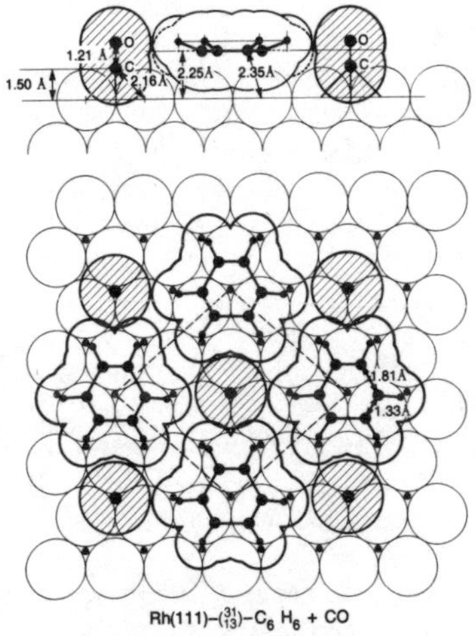

Figure 20: The Rh(111)-$\begin{pmatrix}3&1\\1&3\end{pmatrix}$-$C_6H_6$ + CO LEED Structure.

Table II

Coadsorption of Adsorbates on Rh(111)

Adsorbates	Ratio	Coadsorbed LEED Pattern	Ref.
NO + -C.CH_3	1:1	c(4x2)	96
CO + C_2H_2	1:1	c(4x2)	97
CO + -C.CH_3	1:1	c(4x2)	96
CO + C_6H_6	2:1	(3x3)	94
CO + C_6H_6	1:1	c($2\sqrt{3}$x4)rect	95
CO + C_6H_5F	2:1	(3x3)	97
CO + C_6H_5F	1:1	c($2\sqrt{3}$x4)rect	97
CO + Na	1:1	c(4x2)	98
CO + NO		Disorder	99
Na + C_2H_2		Disorder	99
Na + -C.CH_3		Disorder	99
Na + C_6H_6		Mixed*	99

* - 2 patterns characteristic of individual adsorbates observed suggesting phase separation into independent domains.

is a strong electron donor to metal surfaces. Apparently, therefore, the presence of electron acceptor-donor interactions induces ordering and the formation of surface structures containing both benzene and CO molecules in the same unit cell.

This is not an isolated phenomenon: Table II gives examples of several systems where the coadsorption of an electron donor and an acceptor leads to formation of ordered structures while the coadsorption of two electron donors or two electron acceptors yields disordered surface monolayers. Thus, in these systems at least, it is clear that the attractive forces arising from donor-acceptor interaction are crucially important in determining the stability and structure of the coadsorption system. In the case of the coadsorption of benzene with CO on Rh(111) there is little change in the decomposition/desorption temperatures of either the CO or benzene[93]. By contrast, the coadsorption of CO with alkali metals can have a dramatic influence on the CO binding strength. For example, CO desorbs completely from a clean Cu(110) surface at temperatures below 200K whereas in the presence of coadsorbed potassium two new binding sites are populated yielding CO desorption at 480K and 550K[100]. This corresponds to an increase in the heat of adsorption from around 45kJ/mol to >110kJ/mol. Any such phenomena occurring under catalytic conditions will, of course, play an important role in the reaction concerned and coadsorbed molecules are often used as bonding modifiers, as will be shown later, to change the activity and selectivity of catalysts.

4.5 Surface Dynamics - Adsorption, Diffusion and Reaction

The next concept concerns the dynamics of molecules on surfaces and is sometimes known as the **two dimensional phase approximation**. The basis of the approximation is that the activation energies for diffusion of any adsorbed molecule across a surface are substantially less than the large potential barriers for desorption or, indeed, diffusion into the bulk. It is commonly assumed therefore that at all normal temperatures the adsorbed atoms and molecules can visit all the surface sites within their residence time on the surface and are in equilibrium with each other at the various surface sites. This, for example, explains why attractive inter-adsorbate interactions can lead to the formation of islands of ordered close-packed structures even at submonolayer coverages. At high coverages[101] and in the presence of certain coadsorbates (see above, and ref.102) the mobility of the species will clearly be reduced; nevertheless, this two dimensional phase approximation is used when developing theories of evaporation[103] or crystal growth[104–106] and has been very useful in modelling many catalytic reactions.

A closely related phenomenon that is very important in heterogeneous catalysis is that of "spillover" of adsorbed species[107,108]. In a multiphasic system such as a supported metal catalyst it is possible for molecules to adsorb, and perhaps even decompose or react, on one component before diffusing over onto a second phase where they may react with a different adsorbed species. This concept underlies the principle of operation of many bifunctional catalysts and has been well demonstrated in surface studies on model systems[109–111].

In molecular beam surface scattering studies, it is possible to separately determine the energy accommodation coefficients for translation, rotation and vibration for molecules incident on a surface by monitoring the kinetic, rotational and vibrational energies of both incident and scattered molecules[112]; a set of results for the scattering of NO from a Pt(111) crystal surface[113] is shown in Fig.21. From such experiments it is apparent that most of these modes equilibrate quite well upon a single collision with the surface, thus giving rise to the concept of **rapid gas-surface energy transfer**. This explains why the desorbed products of even the most exothermic reactions are 'cold'. Nevertheless, the accommodation of a molecule on a surface is not a simple process and surface science studies have

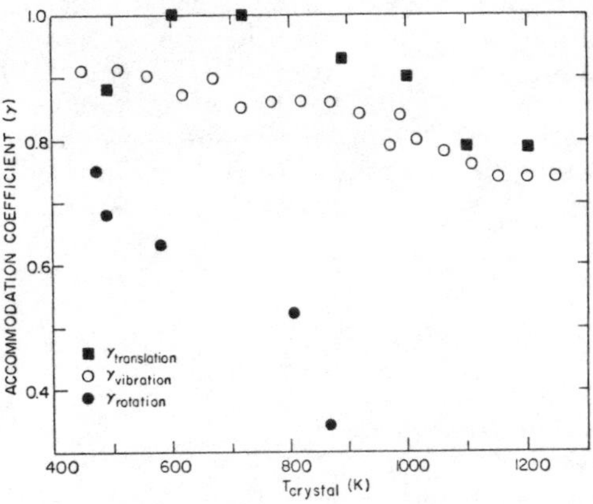

Figure 21: Translational, Rotational and Vibrational Accomodation during the Scattering of NO from Pt(111).

given rise to a further concept, namely that of the **precursor state.** It is often proposed that molecules incident on a surface go into a weakly bound state where they spend a residence time that may amount to hundreds of vibrational oscillations before they either desorb or enter into a more stable, strongly chemisorbed state. Furthermore, such precursor states may be subdivided into two categories - namely, "extrinsic" states in which the molecule is physisorbed above a chemisorbed overlayer and "intrinsic" states in which the precursor is weakly bound to the clean surface itself. The presence of these precursor states has been deduced from atomic and molecular beam scattering experiments as well as desorption and sticking probability studies for many adsorbate surface systems[114,115]. The existence of a second layer physisorbed state has also been well documented by spectroscopic characterization during low temperature adsorption studies: direct and unambiguous spectroscopic characterization of intrinsic precursor states, however, has only recently been reported[116,117]. It should be noted that the term precursor state as used here to describe a weakly bound state which is a precursor to a chemisorbed complex should not be confused with usage relating to the transition from a strongly chemisorbed molecular state to a dissociated one (eg ref.118).

5. Concepts in Heterogeneous Catalysis

Let us now turn to concepts that come directly from studies of catalytic reactions on low area model systems. Useful catalytic processes require a rapid turnover, ie. adsorption, surface diffusion, chemical rearrangement and reaction, and product desorption must all occur in such a manner that the surface can rapidly accommodate new molecules to continue the catalytic conversion. This criterion requires the formation of sufficiently strong chemical bonds between the reactant molecule and substrate to permit bond activation but not so strong as to inhibit interaction with other adsorbed species. The condition is also well illustrated experimentally in the "volcano" shaped plots of activity versus heat of adsorption which are widely found throughout heterogeneous catalysis[119]. Furthermore the binding of the products must not be so strong that the products do not readily desorb since this would lead to stoichiometric as opposed to catalytic reaction.

The first concept to come from studies on well defined surfaces is the existence of two classes of reactions: those that are **structure sensitive** and those that are **structure**

insensitive. Perhaps one of the simplest conceptual reactions is the exchange of hydrogen and deuterium to form HD. This may be studied at low pressures using a mixed molecular beam apparatus in which the H_2 and D_2 are incident on a surface and the HD concentration in the desorbed beam is monitored. Such reactive scattering studies on Pt substrates indicate that atomic steps are the most efficient site for dihydrogen dissociation thereby leading to HD formation[120]. Thus, the reactivity of the close packed (111) surface is about an order of magnitude lower than more open or stepped surfaces since on this face the surface defects are predominantly responsible for hydrogen dissociation[121]. This, therefore, is an example of a structure sensitive reaction and one in which the atomic steps play a unique role.

Table III

Surface Science - High Pressure Studies of Catalytic Systems

Hydrogenation of Ethylene (Pt, Rh)	Ethylene Partial Oxidation (Ag)
Hydrogenation of Carbon Monoxide (Ni, Fe, Rh, Re, Cu, alloys)	Hydrogenation of Benzene, Cyclohexene (Pt, Pd, Rh, alloys)
Oxidation of Carbon Monoxide (Pt)	Hydrodesulfurization of Thiophene (Mo, Re)
Ammonia Synthesis (Fe, Re)	Ammonolysis of Butylalcohol (Rh, Cu)
Ammonia Oxidation (Pt)	Hydrogenolysis of Ethane (Pt, Rh)
Alkane Rearrangements (Pt, Pd, alloys) (Isomerization, Dehydrocyclization & Hydrogenolysis)	Steam Gasification of Carbon (Ni, K)
	Water Gas Shift Reaction (Cu)
Methanol Partial Oxidation (Mo)	Methane Decomposition (Ni, Rh, Ir)

Many other catalytic reactions have now been studied by modern surface science techniques, some of which are listed in Table III; a few of these examples will now be reviewed in greater detail.

The synthesis of ammonia has been studied over various single crystal surfaces of iron. This is a particularly surface structure sensitive reaction - the (111) and (211) surface orientations are about an order of magnitude more active than the (100) & (210) faces and two orders of magnitude more active than the close packed (110) face; this latter surface being the least active of all those studied[122](Fig.22). Early studies by Boudart[123] led to the suggestion that the seven-fold coordination (C_7) site that is present only on certain iron surfaces is the most active for the dissociation of dinitrogen to atomic nitrogen - the postulated rate determining step in ammonia synthesis. The studies in our own laboratory on the (111) and (211) faces showing these to be the most active in ammonia synthesis seem to confirm this hypothesis since only these two single crystal surfaces contain the C_7 Fe sites in the second layer of the surface (Fig.23).

A somewhat more complicated example of structure sensitivity and insensitivity is the hydrodesulfurization reaction; a very important process used to remove sulphur from an oil feed. This reaction may be modelled by the hydrodesulfurization of thiophene to butane, butenes and butadiene. When this reaction is carried out on molybdenum and rhenium single crystal surfaces it exhibits structure insensitivity over molybdenum but significant structure sensitivity over rhenium[124](Fig.24). Furthermore, in agreement with results on high surface area materials[125], the specific rates over the rhenium single crystals are 1-5 times higher than over molybdenum. This appears to result from the presence of a stable carbonaceous and/or sulphur overlayer on molybdenum surfaces which not only

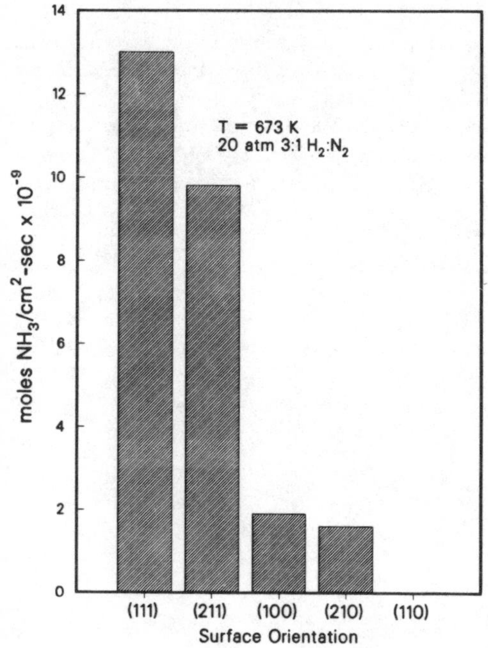

Figure 22: Structure Sensitivity of Ammonia Synthesis over Iron Single Crystals.

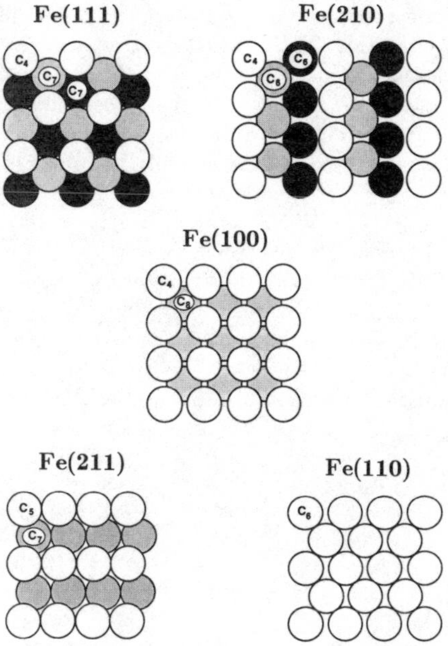

Figure 23: Hard Sphere Models of bcc-Fe Surfaces showing Surface Atom Coordination.

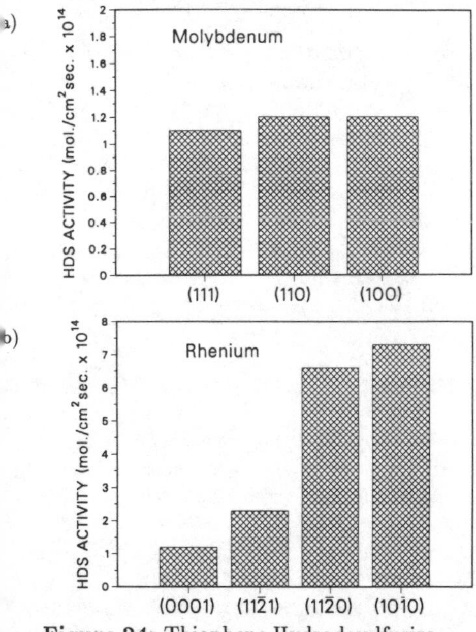

Figure 24: Thiophene Hydrodesulfurization over Single Crystal Surfaces.

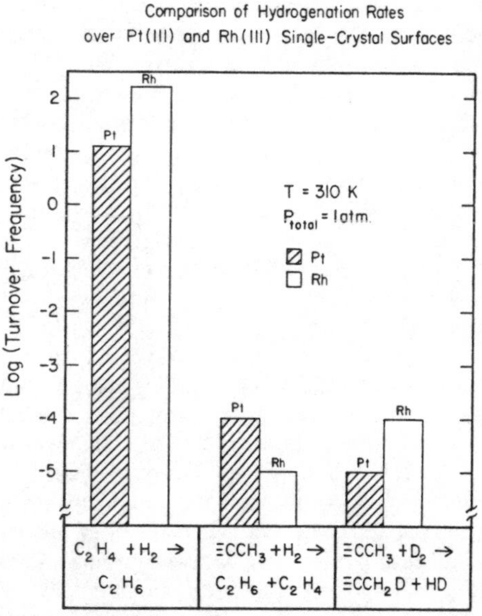

Figure 25: Hydrogenation Rates over Pt(111) and Rh(111) Single Crystal Surfaces.

moderates the highly active surface but also masks the surface structure sensitivity[126,127]. In contrast, rhenium surfaces remain free of irreversibly bound sulphur and carbon under reaction conditions and the different electronic and geometrical properties of the different crystal faces give rise to the observed structure sensitivity[128].

An example of a structure insensitive reaction is ethylene hydrogenation at low temperatures (~300K). This reaction has been extensively studied on Pt(111) and Rh(111) single crystals[129–133]: under these conditions (310K/1 atm.) the metal surfaces are completely covered with a stable adsorbed layer of ethylidyne. The rehydrogenation of this species, and indeed the exchange rate of deuterium into the methyl group, is many orders of magnitude slower than the ethylene hydrogenation rate (Fig.25). The reaction itself appears to occur on top of this ethylidyne overlayer and the purpose of the metal is secondary; more specifically, to dissociate the molecular hydrogen. The ethylidyne deposit can then transfer the hydrogen to weakly bound ethylene adsorbed on top of it resulting in hydrogenation and the formation of ethane (Fig.26).

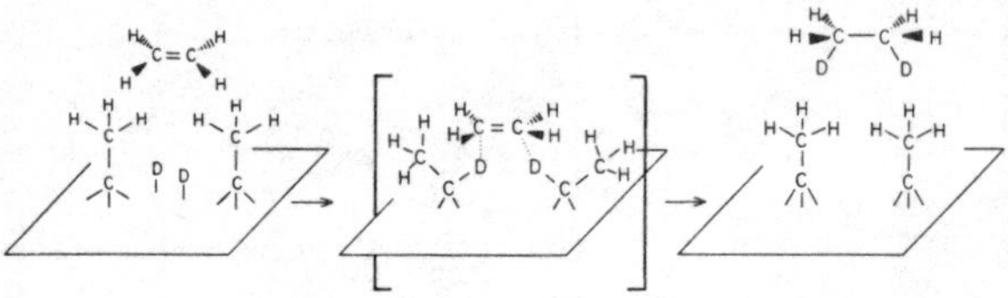

Figure 26: Proposed Mechanism for Ethylene Hydrogenation at Low Temperatures.

Another concept in catalysis is the use of **bonding and structural modifiers** - collectively known as 'promoters' - to change the catalytic activity and selectivity. A classic example of a structural modifier is that of alumina in ammonia synthesis over iron catalysts. Model studies have shown that when alumina is added in the form of islands to the inactive Fe(110) surface and the system then heated in water vapor, the ensuing oxidation of the iron is accompanied by migration onto the alumina and substantial restructuring results[134] - various authors have also proposed the formation of an iron aluminate ($FeAl_2O_4$)[135,136]. Subsequent reduction under reaction conditions yields metallic iron crystallites in orientations that are very much more active (Fig.27) than the original surface (ie (111) and (211) as opposed to (110)). The primarily role of the alumina is to stabilise the highly active restructured surface produced by the hydrothermal treatment since transient restructuring and enhanced activity is seen after such treatment even in the absence of alumina. The effect is not restricted to the (110) surface; other inactive surfaces of iron may also be converted to ones containing highly active (111) or (211) crystal faces in the presence of alumina.

Alkali metals are extensively used as promoters in commercial catalyst formulations[137]. The dramatic effect that coadsorption of potassium can have on the strength of molecular CO chemisorption has already been mentioned. A similar increase in binding strength is also observed in the CO/Rh(111) system: furthermore, isotopic scrambling occurs for $\theta_K \geq 0.1$ but not on the K-free Rh(111) surface[138](Fig.28). This is indicative of alkali induced cleavage of the C–O bond and an example of the profound effects that modifiers may have on the chemical bonding of adsorbates.

Similar explanations have been put forward to explain the promoting effect of potassium in ammonia synthesis, ie that the alkali enhances dinitrogen dissociation. Recent work, however, suggests that the primary role of the potassium is to alleviate product inhibition of the reaction[139]. At high conversions (ie high ammonia partial pressures) active sites for the ammonia synthesis are blocked by adsorbed product molecules. Coadsorbed

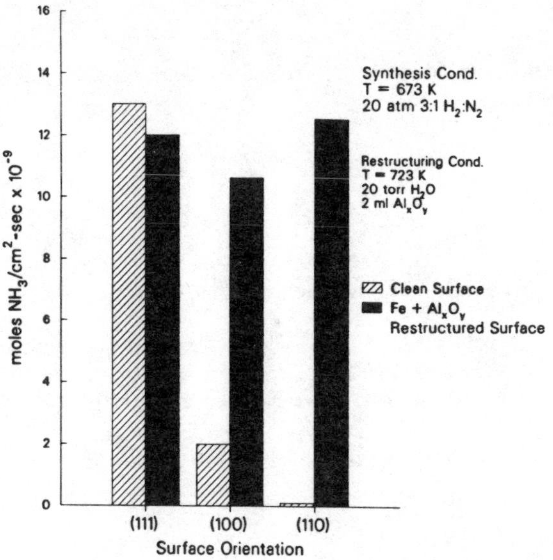

Figure 27: The Effect of Alumina-Induced Reconstruction of Iron Surfaces on the Rate of Ammonia Synthesis.

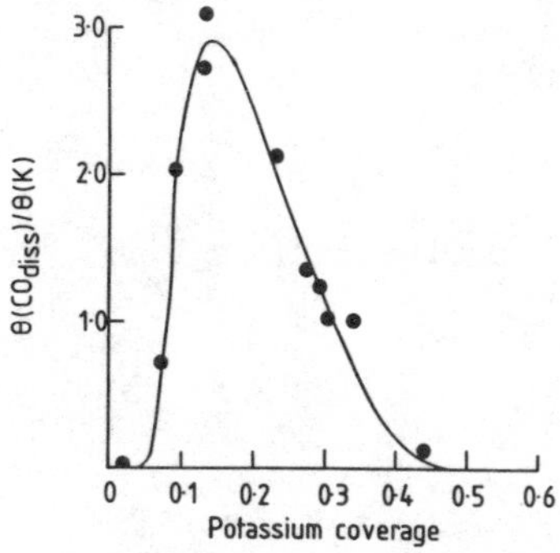

Figure 28: Number of CO Molecules dissociated per Potassium Atom as a function of Potassium Coverage on Rh(111) (as deduced from $^{13}C^{16}O/^{12}C^{18}O$ scrambling experiments).

potassium weakens the bonding of the ammonia leading to a lower steady state surface concentration and, hence, increased activity. Thus alkalis may not only promote reactions by activation of the reactant molecules but also by a weakening of the interaction of the product molecules with the surface.

6. Summary

The application of modern surface science techniques has revolutionized our understanding of phenomena occurring at the gas-solid interface and, as we have seen above, led to a number of well defined new concepts. In contrast, progress in developing and applying surface sensitive technology to study other interphasic boundaries (eg gas-liquid and liquid-solid interfaces) has been much slower and this must undoubtedly be one of the greatest challenges that lie ahead. This would open up the way for *in situ* molecular level studies in areas such as electrochemistry, biological surfaces, tribology and corrosion.

Within the field of catalysis, it is also clear that the application of existing surface science techniques and concepts has had a profound influence on the way in which we view the fundamental steps that underly catalytic processes. Whilst much remains to be discovered, and still more subtle nuances lie uncovered, our molecular level understanding of surface and catalytic phenomena has now reached a point where its application may begin to be used to help develop new generations of high technology catalysts. Certainly there are a vast number of areas which could benefit greatly by such developments ; these include chemical energy conversion, low temperature catalysis, pharmaceutical production and the synthesis of complex organic systems. It is the concept of catalyst design, therefore, that is perhaps the most important area in the future of catalytic science.

Acknowledgements

This work was supported by the Director, Office of Basic Energy Science, Materials Science Division of the U.S. Department of Energy.

References

(1) Rabo, J.A., Ed. "Zeolite Chemistry and Catalysis", *Am. Chem. Soc. Monogr.* **1976**, 171.
(2) Vaughan, D.E.W.; Chen, N.Y. and Degnan, T.F. in *Chemical Engineering Progress* (Feb) **1988**.
(3) Flanigen, E.M.; Lok, B.M.; Patton, R.L.; Wilson, S.T. in "New Developments in Zeolite Science and Technology", *Stud. Surf. Sci. Catal.* 28, 103 (Elsevier, Amsterdam, 1986).
(4) Woodruff, D.P.; Delchar, T.A., "Modern Techniques in Surface Science" (CUP, Cambridge, 1986).
(5) Blakely, D.W.; Kozak, E.I.; Sexton, B.A.; Somorjai, G.A., *J. Vac. Sci. Tech.* **1976**, 13, 1091.
(6) Goodman, D.W., *Acc. Chem. Res.* **1984**, 17, 194.
(7) Campbell, C.T.; Paffett, M.T., *Surf. Sci.* **1984**, 139, 396.
(8) Rucker, T.G.; Frank, K.; Colomb, D.; Logan, M.A.; Somorjai, G.A., *Rev. Sci. Instrum.* **1987**, 58, 2292.
(9) Somorjai, G.A., "Chemistry in Two Dimensions: Surfaces" (Cornell University Press, 1981).
(10) Gibbs, J.W.,"The Collected Works of J. Willard Gibbs", Vol.1 (Yale University Press, New Haven, 1948).
(11) Kelley, M., *J. Catal.* **1979**, 57, 113.
(12) Abraham, F.F.; Brundle, C.R., *J. Vac. Sci. Tech.* **1981**, 18, 506.
(13) Miedema, A.R., *Z. Metallk.* **1978**, 69, 455.
(14) Chelikowsky, J.R., *Surf. Sci.* **1984**, 139, L197.
(15) Lang, N.D.; Kohn, W., *Phys. Rev.* **1970**, B1, 4555 and **1971** B3, 1215.
(16) Defay, R.; Prigogine, I.; Bellemans, A.; Everett, D.H., "Surface Tension and Adsorption" (Longmans, New York, 1966).
(17) Borel, J.-P.; Chatelain, A., *Surf. Sci.* **1985**, 156, 572 and refs. therein.
(18) MacLaren, J.M.; Pendry, J.B.; Rous, P.J.; Saldin, D.K.; Somorjai, G.A.; Van Hove M.A.; Vvedensky, D.D., "Surface Crystallographic Information Service - A Handbook of Surface Structures" (D. Reidel Publ., Dordrecht, 1987).
(19) Van Hove, M.A.; Koestner, R.J.; Stair, P.C.; Biberian, J.P.; Kesmodel, L.L.; Bartos, I.; Somorjai, G.A., *Surf. Sci.* **1981**, 103, 189 & 218.
(20) Tromp, R.M.; Smeenk, R.G.; Saris, F.W.; Chadi, D.J., *Surf. Sci.* **1983**, 133, 137.
(21) Holland, B.W.; Duke, C.B.; Paton, A., *Surf. Sci.* **1984**, 140, L269.
(22) Hsu, T., *Ultramicroscopy* **1983**, 11, 167.
(23) Bauer, E., *Ultramicroscopy* **1985**, 17, 51.
(24) Smith, D.J., *Surf. Sci.* **1986**, 178, 462.
(25) Binnig, G.; Rohrer, H., *IBM J. Res. Develop.* **1986**, 30, 355.
(26) Harsma, P.K.; Tersoff, J., *J. Appl. Phys.* **1987**, 61, R1.
(27) Marchon, B.; Ogletree, D.F.; Salmeron, M.; Siekhaus, W., *J. Vac. Sci. Tech.* **1988**, A6, 531.
(28) Besocke, K.; Wagner, H., *Surf. Sci.* **1975**, 52, 653 & *Phys. Rev.* **1973**, B8, 4597.
(29) Bernasek, S.L.; Somorjai, G.A., *J. Chem. Phys.* **1975**, 62, 3149.
(30) Blakely, D.W.; Somorjai, G.A., *Surf. Sci.* **1977**, 65, 419.

(31) Wandelt, K.; Hulse, J.; Kuppers, J., *Surf. Sci.* **1981**, 104, 212 .
(32) Siddiqui, H.R.; Guo, X.; Chorkendorff, I.; Yates, J.T., *Surf. Sci.* **1987**, 191, L813.
(33) Iijima, S.; Ichihashi, T., *Phys. Rev. Letts.* **1986**, 56, 616.
(34) Pettiette, C.L.; Yang, S.H.; Craycraft, M.J.; Conceicao, J.; Laaksonen, R.T.; Cheshnovsky, O.; Smalley, R.E., *J. Chem. Phys.* **1988**, 88, 5377.
(35) Bouffat, P.; Borel, J.P., *Phys. Rev.* **1976**, A13, 2287.
(36) King, D.A.; Thomas, G., *Surf. Sci.* **1980**, 92, 201.
(37) Barker, R.A.; Estrup, P.J.; Jona, F.; Marcus, P.M., *Solid State Comm.* **1978**, 25, 375.
(38) Onuferko, J.H.; Woodruff, D.P.; Holland, B.W., *Surf. Sci.* **1979**, 87, 357.
(39) Razon, L.F.; Schmitz, R.A.; *Cat. Rev.- Sci. Eng.* **1986**, 28, 89.
(40) Imbihl, R.; Cox, M.P.; Ertl, G., *J. Chem. Phys.* **1986**, 84, 3519 and refs. therein.
(41) Yeates, R.C., Turner, J.E.; Gellman, A.J.; Somorjai, G.A., *Surf. Sci.* **1985**, 149, 175.
(42) Collins, N.A.; Sundaresan, S.; Chabal, Y.J., *Surf. Sci.* **1987**, 180, 136.
(43) Schwartz, S.B.; Schmidt, L.D., *Surf. Sci.* **1987**, 183, L269.
(44) Ladas, S.; Imbihl, R.; Ertl, G., *Surf. Sci.* **1988**, 197, 153 and refs. therein.
(45) Yao, H.C.; Sieg, M.; Plummer, H.K., *J. Catal.* **1979**, 59, 365.
(46) Wang, T.; Schmidt, L., *J. Catal.* **1981**, 70, 187.
(47) Nakayama, T,; Arai, M.; Nishiyama, Y., *J. Catal.* **1983**, 79, 497.
(48) Rous, P.J.; Pendry, J.B.; Saldin, D.K.; Heinz, K.; Muller, K.; Bickel, N., *Phys. Rev. Lett.* **1986**, 57, 2951.
(49) Schmidt, L.D.; Wang, T., *J. Vac. Sci. Tech.* **1981**, 18, 520.
(50) Hoflund, G.B.; Asbury, D.A.; Kirszensztejn, P.; Laitinen, H.A., *Surf. Sci.* **1985**, 161, L583.
(51) Vanini, F.; Buchler, St.; Yu, X.; Erbudak, M.; Schlapbach, L.; Baiker, A., *Surf. Sci.* **1987**, 189/190, 1117 and refs. therein.
(52) Bouwman, R.; Lippits, G.J.M.; Sachtler, W.M.H., *J. Catal.* **1972**, 25, 350.
(53) Tomanek, D.; Mukherjee, S.; Kumar, V.; Bennemann, K.H., *Surf. Sci.* **1982**, 114, 11.
(54) Munuera, G.; Gonzalez-Elipe, A.R.; Espinos, J.P.; Conesa, J.C.; Soria, J.; Sanz, J., *J. Phys. Chem.* **1987**, 91, 6625.
(55) Gorte, R.J.; Altmann, E.; Corallo, G.R.; Davidson, M.R.; Asbury, D.A.; Hoflund, G.B., *Surf. Sci.* **1987**, 188, 327.
(56) Hoflund, G.B.; Grogan, A.L.; Asbury, D.A., *J. Catal.* **1988**, 109, 226.
(57) Bauer, E., in "The Chemical Physics of Solid Surfaces and Heterogeneous Catalysis" (Eds. King, D.A.; Woodruff, D.P.) **1984**, Vol.3.
(58) Ohtani, H.; Kao, C.T.; Van Hove, M.A.; Somorjai, G.A., *Progr. Surf. Sci.* **1987**, 23, 155.
(59) Binns, C.; Norris, C.; Smith, G.C.; Padmore, H.A.; Barthes-Labrousse, M.G., *Surf. Sci.* **1983**, 126, 258.
(60) Clarke, A.; Rous, P.J.; Arnott, M.; Jennings, G.; Willis, R.F., *Surf. Sci.* **1987**, 192, L843 and refs. therein.
(61) Mehta, S.; Simmons, G.W.; Klier, K.; Hermann, R.G., *J. Catal.* **1979**, 57, 339.
(62) Sinfelt, J.H.; Via, G.H.; Lytle, F.W., *Catal. Rev.- Sci. Eng.* **1984**, 26, 81.
(63) Groomes, D.O.; Wynblatt, P., *Surf. Sci.* **1985**, 160, 475 and refs. therein.
(64) Baird, R.J.; Ogletree, D.F.; Van Hove, M.A.; Somorjai, G.A., *Surf. Sci.* **1986**, 165, 345.
(65) Vurens, G.H.; Salmeron, M.; Somorjai, G.A., *Surf. Sci.*, in press (1988).
(66) Badyal, J.P.S.; Gellmann, A.J.; Judd, R.W.; Lambert, R.M., *Catal. Letts.* **1988**, 1, 41.

(67) Zhou, Y.; Nakashima, M.; White, J.M., *J. Phys. Chem.* **1988** 92, 812.
(68) Pretorius, R.; Harris, J.M.; Nicolet, M.A., *Solid State Electron.* **1978**, 2, 667.
(69) Kung, M.C.; Kung, H.H., *Surf. Sci.* **1981**, 104, 253.
(70) Vurens, G.; Strongin, D.R.; Salmeron, M.; Somorjai, G.A., *Surf. Sci.* **1988**, 199, L387.
(71) Bagus, P.S.; Hermann, K.; Seel, M., *J. Vac. Sci. Tech.* **1981**, 18, 435.
(72) Hermann, K.; Hass, H.J., *Surf. Sci.* **1987**, 189/190, 426 and refs. therein.
(73) Ohtani, H.; Van Hove, M.A.; Somorjai, G.A., *ICSOS Proc.* **1987**.
(74) Gomez-Sal, M.P.; Johnson, B.F.G.; Lewis, J.; Raithby, P.R.; Wright, A.H., *J.Chem. Soc.- Chem. Comm.* **1985**, 1682.
(75) Connoly, M.; Somers, J.; Bridge, M.E.; Lloyd, D.R., *Surf. Sci.* **1987**, 185, 559 and refs. therein.
(76) Mate, C.M.; Somorjai, G.A.; Tom, H.W.K.; Zhu, X.D.; Shen, Y.R., *J. Chem. Phys.* **1988**, 88, 441.
(77) DiNardo, N.J.; Avouris, Ph.; Demuth, J.E., *J. Chem. Phys.* **1984**, 81, 2169.
(78) Grassian, V.H.; Muetterties, E.L., *J. Phys. Chem.* **1986**, 90, 5900.
(79) Bader, M.; Hasse, J.; Frank, K.-H.; Puschmann, A.; Otto, A., *Phys. Rev. Letts.* **1986**, 56, 1921 and refs. therein.
(80) Gurney, B.A.; Ho, W., *J. Chem. Phys.* **1987**, 87, 1376.
(81) Koel, B.E.; Crowell, J.E.; Bent, B.E.; Mate, C.M.; Somorjai, G.A., *J. Phys. Chem.* **1986**, 90, 2949.
(82) Koestner, R.J.; Van Hove, M.A.; Somorjai, G.A., *Surf. Sci.* **1982**, 121, 321.
(83) Bent, B.E.; Mate, C.M.; Kao, C.T.; Slavin, A.L.; Somorjai, G.A., *J. Phys. Chem.*, in press (1988).
(84) Mate, C.M.; Kao, C.T.; Bent, B.E.; Somorjai, G.A., *Surf. Sci.*, in press (1988).
(85) Van Hove, M.A.; Bent, B.E.; Somorjai, G.A., *J. Phys. Chem.* **1988**, 92, 973.
(86) Bent, B.E.; Mate, C.M.; Crowell, J.E.; Koel, B.E.; Somorjai, G.A., *J. Phys. Chem.* **1987**, 91, 1493.
(87) Kang, D.B.; Anderson, A.B., *Surf. Sci.* **1985**, 155, 639.
(88) Godbey, D.; Zaera, F.; Yeates, R.; Somorjai, G.A., *Surf. Sci.* **1986**, 167, 150.
(89) Ishi, S.; Ohno, Y.; Viswanathan, B., *Surf. Sci.* **1985**, 161, 349 and refs. therein.
(90) Tracy, J.C.; Palmberg, P.W., *J. Chem. Phys.* **1969**, 51, 4852.
(91) Ogletree, D.F.; Van Hove, M.A.; Somorjai, G.A., *Surf. Sci.* **1986**, 173, 351.
(92) Van Hove, M.A.; Koestner, R.J.; Frost, J.C.; Somorjai, G.A., *Surf. Sci.* **1983**, 129, 482.
(93) Mate, C.M.; Somorjai, G.A., *Surf. Sci.* **1985**, 160, 542.
(94) Blackman, G.S.; Lin, R.F.; Van Hove, M.A.; Somorjai, G.A., *Acta Cryst.* **1987**, B43, 368.
(95) Van Hove, M.A.; Lin, R.F.; Somorjai, G.A., *J. Am. Chem. Soc.* **1986**, 108, 2532.
(96) Blackman, G.S.; Kao, C.T.; Bent, B.E.; Mate, C.M.; Van Hove, M.A.; Somorjai, G.A., *Surf. Sci.*, submitted (1988).
(97) Mate, C.M., Ph.D. Thesis (Berkeley) **1986**.
(98) Mate, C.M.; Bent, B.E.; Somorjai, G.A., *J. Elec. Spec. Relat. Phenom.* **1986**, 39, 205.
(99) Kao, C.T.; Mate, C.M.; Blackman, G.S.; Bent, B.E.; Van Hove, M.A.; Somorjai, G.A., *J. Vac. Sci. Tech.*, in press (1988) and ref.97.
(100) Lackey, D.; Surman, M.; Jacobs, S.; Grider, D.; King, D.A., *Surf. Sci.* **1985**, 152/153, 513.
(101) Butz, R.; Wagner, H., *Surf. Sci.* **1977**, 63, 448.
(102) Zhadanov, V.P., *Surf. Sci.* **1988**, 194, 1.

(103) Hirth, J.P.; Pound, A.M., "Condensation and Evaporation" (Pergamon Press, Oxford, 1963).
(104) Burton, W.K.; Cabrera, N.; Frank, N.C., *Phil. Trans. Roy. Soc. (Lond.)* **1951**, 243A, 299.
(105) DeMiguel, J.J.; Sanchez, A.; Cebollada, A.; Gallego, J.M.; Feiron, J.; Ferrer, S., *Surf. Sci.* **1987**, 189/190, 1062.
(106) Joyce, B.A.; Dobson, P.J.; Neave, J.H.; Zhang, J., *Surf. Sci.* **1986**, 178, 110 and refs. therein.
(107) Pajonk, G.M.; Teichner, S.J.; Germain, J.E. Eds. *Studies in Surface Science and Catalysis 17*: "Spillover of Adsorbed Species" (Elsevier, Amsterdam, 1983).
(108) Connor, W.C.; Pajonk, G.M.; Teichner, S.J., *Adv. Catal.* **1986**, 34, 1.
(109) Erre, R.; Fripiat, J.J., in ref.107.
(110) Goodman, D.W.; Yates, J.T.; Peden, C.H.F., *Surf. Sci.* **1985**, 164, 417.
(111) Cevallos Candau, J.F.; Conner, W.C., *J. Catal.* **1987**, 106, 378.
(112) Hager, J.; Shen, Y.R.; Walther, H., *Phys. Rev.* **1985**, A31, 1962 and refs. therein.
(113) Asscher, M.; Guthrie, W.L.; Lin, T.H.; Somorjai, G.A., *J. Chem. Phys.* **1983**, 78, 6992.
(114) Barker, J.A.; Auerbach, D.A., *Surf. Sci. Rept.* **1985**, 4, 1.
(115) D'Evelyn, M.P.; Steinruck, H.-P.; Madix, R.J., *Surf. Sci.* **1987**, 180, 47.
(116) Norton, P.R.; Tapping, W.L.; Goodale, J.W., *Surf. Sci.* **1978**, 72, 33.
(117) Prince, K.C.; Paolucci, G.; Bradshaw, A.M., *Surf. Sci.* **1986**, 175, 101.
(118) Moon, D.W.; Cameron, S.; Zaera, F.; Eberhardt, W.; Carr, R.; Bernasek, S.L.; Gland, J.L.; Dwyer, D.J., *Surf. Sci.* **1987**, 180, L123.
(119) Gasser, R.P.H., "An Introduction to Chemistry and Catalysis by Metals" (Clarendon, Oxford, 1985).
(120) Salmeron, M.; Gale, R.J.; Somorjai, G.A., *J. Chem. Phys.* **1977**, 67, 5324 & **1979**, 70, 2807.
(121) Poelsema, B.; Verheij, L.K.; Comsa, G., ECOSS-6, **1984**.
(122) Strongin, D.R.; Carrazza, J.; Bare, S.R. Somorjai, G.A., *J. Catal.* **1987**, 103, 213.
(123) Dumesic, J.A.; Topsoe, H.; Boudart, M., *J. Catal.* **1975**, 37, 513.
(124) Bussell, M.E.; Gellmann, A.J.; Somorjai, G.A., *J. Catal.* **1988**, 110, 423.
(125) LeDoux, M.J.; Michaux, O.; Agostini, G.; Panissod, P., *J. Catal.* **1986**, 102, 275 and refs. therein.
(126) Bussell, M.E.; Somorjai, G.A., *J. Catal.* **1987**, 106, 93.
(127) Gellman, A.J.; Bussell, M.E.; Somorjai, G.A., *J. Catal.* **1987**, 107, 103.
(128) Bussell, M.E.; Somorjai, G.A., *J. Phys. Chem.*, in press (1988).
(129) Zaera, F.; Somorjai, G.A., *J. Am. Chem. Soc.* **1984**, 106, 2288 .
(130) Koel, B.E.; Bent, B.E.; Somorjai, G.A., *Surf. Sci.* **1984**, 146, 211.
(131) Davis, S.M.; Zaera, F.; Gordon, B.E.; Somorjai, G.A., *J. Catal.* **1985**, 92, 240.
(132) Zaera, F.; Somorjai, G.A., *J. Phys. Chem.* **1985**, 89, 3211.
(133) Godbey, D.; Zaera, F.; Yeates, R.; Somorjai, G.A., *Surf. Sci.* **1986**, 167, 150.
(134) Strongin, D.R.; Bare, S.R.; Somorjai, G.A., *J. Catal.* **1987**, 103, 289.
(135) Sushumna, I.; Ruckenstein, E., *J. Catal.* **1985**, 94, 239.
(136) Paparazzo, E.; *Appl. Surf. Sci.* **1986**, 25, 1 and refs. therein.
(137) Mross, W.D., *Catal. Rev.- Sci. Eng.* **1983**, 25, 591.
(138) Crowell, J.E.; Tysoe, W.T.; Somorjai, G.A., *J. Phys. Chem.* **1986**, 89, 1598.
(139) Strongin, D.R.; Somorjai, G.A., *J. Catal.* **1988**, 109, 51.

MOLECULAR MECHANISMS OF SPECIFICITY IN DNA-ANTITUMOR DRUG INTERACTIONS

Bernard Pullman

Institut de Biologie Physico-Chimique
Fondation Edmond de Rothschild
13, rue Pierre et Marie Curie
75005 Paris
France

Abstract

One of the central problems in the study of the mechanism of DNA-antitumor drug interactions is the existence and nature of sequence specificity with respect to the base pairs of DNA. Results of a theoretical exploration of this problem are presented for a particularly important group of such drugs, namely the groove binding ligands. The great majority of the investigated antitumor agents of this category show a marked specificity for the minor groove of AT sequences of B-DNA. Contrary to current concepts and some proposals, hydrogen bond formation between the ligands and the receptor bases of DNA or contacts between specific hydrogen atoms on the ligand and on the bases do not appear to be the main determinants of this specificity. The essential factor responsible for this preference is the strong concentration of the electrostatic molecular potential in this groove of these sequences in B-DNA. This distribution of the potential renders difficult the conception of drugs specific for the minor groove of GC sequences and is responsible for the failure in this respect of lexitropsins and isolexins. Nevertheless, for the first time in theoretical computations, a particular isolexin was conceived which should bind preferentially to the minor groove of GC sequence : this is a neutral ligand composed of three furan or imidazole rings linked by NH groups. An increased GC specificity is, moreover, predicted for vinylexins (analogs of isolexins with C=C linkers) and an increased binding energy, with the specificity conserved, may be obtained for monocationic such ligands. In fact, monocationic vinylexins should form a family of universal groove binding ligands, the AT or GC specificity of which may be dictated by the arrangement of their hydrogen bond donor or acceptor heteratomic ring systems.

J. Jortner and B. Pullman (eds.), Perspectives in Quantum Chemistry, 123–144.

I Introduction

Antitumor drugs interact with DNA, their main cellular target, in a number of ways : groove binding, intercalation, covalent bond formation, coordination. One of the central problems in the study of the mechanism of these interactions, common to all of them, is the existence and nature of sequence specificity with respect to the base pairs of DNA. The presence of such a specificity could be of particular significance because it might possibly mean the involvement of specific genes in the effectiveness of the different drugs. The elucidation of the factors responsible for the specificity could then be important for the development of compounds susceptible to contribute to the control of gene expression and also the development of rationally conceived, improved new generations of effective and specific chemotherapeutic agents. Important recent achievements, experimental and theoretical, in the analysis of such sequence specificities open prospects for possible rapid progress in this field. Among the experimental procedures the footprinting and affinity cleaving techniques seem particularly promising in this respect (see e.g. 1,2). So are also the recently developed refined computational techniques for the study of intermolecular interactions involving large biological molecules and polymers, such as AMBER (3), SIBFA (4,5), JUMNA (6,7) and others. A particularly interesting aspect of the latter explorations is that they have brought into evidence the predominant importance in the determination of the sequence selectivity of generally previously unsuspected new factors. It is these factors that I wish to describe here for one of the essential modes of DNA-antitumor drug interactions, which has attracted in recent years great attention, namely, groove binding.

II Sequence specificity in the interaction of "classical" groove binding ligands with DNA

Compounds I-IV constitute some of the most representative and best known ligands of this series. One of the most striking features of their interaction with DNA is their triple specificity for binding to : 1) the minor groove, 2) of AT sequences, 3) of B-DNA (for a recent review see 8). However, while the experimental evidence on this preference is well substantiated, the associated interpretations, based generally on some apparent structural features of the two most studied compounds of the series, netropsin and distamycin A, have given rise to ample discussions. Broadly speaking there are three main proposals in this respect, corresponding to the consideration of three factors as the dominant ones in imposing the above indicated

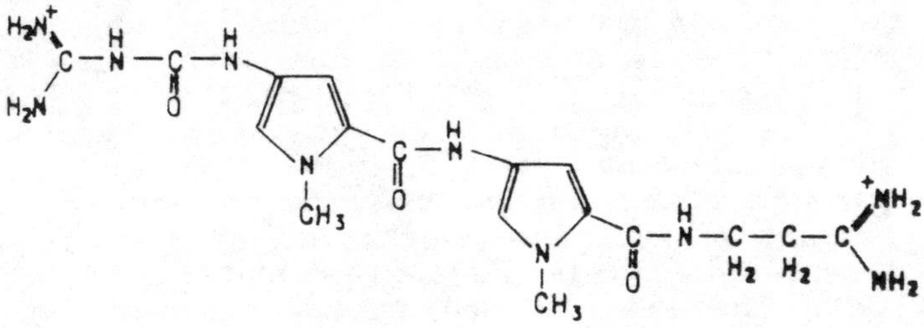

I. Netropsin

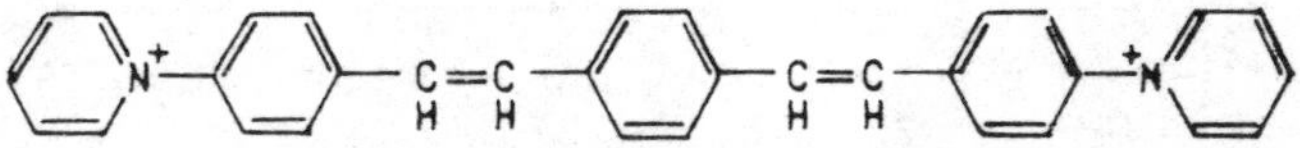

II. SN 18071

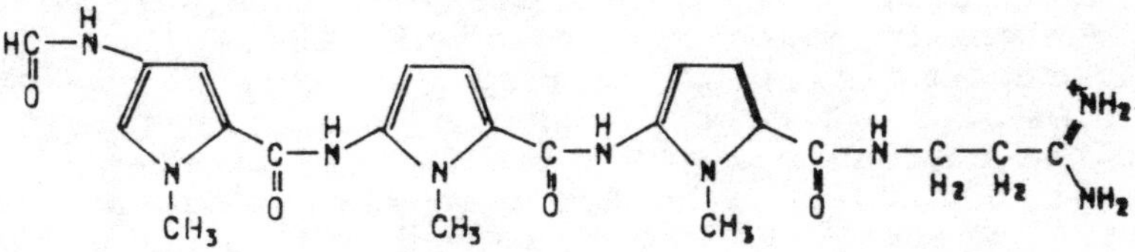

III. Distamycin A (Dist 3)

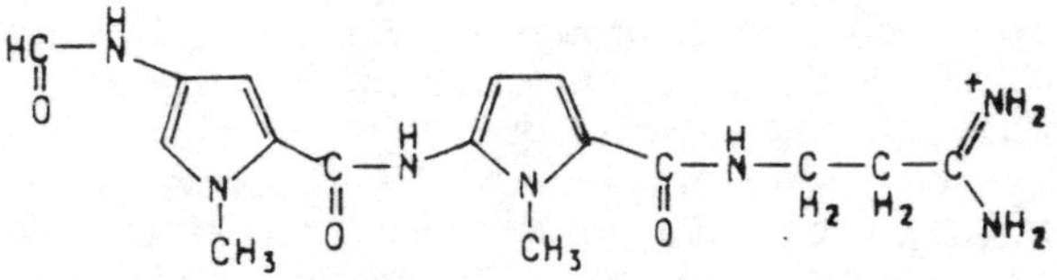

IV. Distamycin 2

specificity.

1) The current and historically first proposal attributes this specificity essentially to hydrogen bond formation between the peptidic NH groups of the drugs and the O2 atoms of thymine and/or N3 atoms of adenine situated in the minor groove of DNA (9,10). This proposition was considered to be strengthened by the bent structure of netropsin, established by X-ray analysis (11) and indicating that the hydrogen donor groups are located on the concave side of the antibiotic and can thus establish an easy contact with the hydrogen accepting atoms in the minor groove of DNA. The charged end groups of the drugs were considered to be also involved in the interaction probably with the phosphate groups of DNA. The reality of the hydrogen bonds is confirmed by an X-ray study of the crystal structure of a complex between netropsin and the double-helical DNA dodecamer CGCGAATT^{Br}CGCG (12) and by NMR studies on the association of netropsin and distamycin A with AT sequences in oligonucleotides (13,14).

That the situation is, however, more complicated than this simple picture suggests and that, in particular, the precise role of the hydrogen bonds in determining specificity has to be reconsidered becomes evident from the examination of other molecules studied. Thus, in particular the bisquaternary ammonium heterocycle SN 18071, which has no hydrogen bonding possibilities, binds also to DNA and shows a similar AT minor groove specificity (15,16). The role of the charged end groups as factors of specificity is also questionable since it has been shown that a netropsin derivative with both ends removed, also complexes to poly(dA).poly(dT) (8).

2) An indication that the source of the specificity common to these diverse drugs (and many others) may reside to a large extent in the properties of the minor groove of AT sequences of B-DNA, rather than in special features of the drugs, was suggested first (17) by the observation that the grooves are the sites of location of the deepest <u>molecular electrostatic potential</u> in DNA, that for AT sequences the deepest potential occurs in their minor groove and that the deepest potentials in DNA are those of the minor groove of AT sequences (18,19). One could conceive then that provided that this groove could offer an appropriate steric fit to the drugs involved, the origin of specificity could reside in the combination of this fit with a corresponding strong electrostatic interaction. The steric fit could be favoured by the inherently greater narrowness of the minor groove of AT regions of DNA in comparison with its GC regions (20).

This hypothesis was rapidly confirmed by explicit computations of interaction energies between a number of compounds of type I-IV with model poly(dA).poly(dT) and poly(dG).poly(dC) duplexes in B-DNA conformation.

Table I

Interaction energies for the formation of DNA-ligand complexes (kcal/mole)

	DNA			Interaction energy			
Ligand	Conformation	Sequence	Groove	Electrostatic	Lennard-Jones	Ligand conformational energy	Total
Netropsin	B	AT	Min.	- 203.5	- 42.3	7.6	- 238.2
			Maj.	- 104.4	- 19.5	15.3	- 108.7
		GC	Min.	- 155.3	- 34.1	8.2	- 181.2
			Maj.	- 138.0	- 15.6	17.2	- 136.4
Distamycin A	B	AT	Min.	- 95.9	- 48.8	15.3	- 129.4
		GC	Min.	- 77.8	- 42.4	18.0	- 102.2
SN 18071	B	AT	Min.	- 86.4	- 31.1	9.8	- 107.7
			Maj.	- 49.8	- 21.3	9.8	- 61.3
		GC	Min.	- 72.2	- 25.8	9.8	- 88.2
			Maj.	- 65.6	- 21.3	9.8	- 77.1

The first such study (21) was performed taking into account the electrostatic and Lennard-Jones components of the interaction energy, the latter one being considered as a measure of the quality of the steric fit between the ligand and the macromolecule. The essential results obtained for the most representative compounds are indicated in Table I. They show, for all the compounds investigated, that the greatest values of the interaction energy are obtained for interaction with the minor groove of the AT sequences. The details of the computations indicate that this preference is favoured both by the electrostatic and the Lennard-Jones energy terms, the first one predominating numerically. This demonstrated that whatever the significance of hydrogen bonds for the stability of the complex, the formation of these bonds is not necessary neither for binding nor for the preference for the minor groove of the AT sequences of B-DNA. It seems, in conformity with the original hypothesis, that if a relatively good steric fit can be obtained in the minor groove the ligand will be sufficiently stabilized there by the favourable electrostatic potentials generated by the AT sequences.

When possible, hydrogen bonds between the proton donating sites of the ligand and the proton accepting sites of the macromolecule are, of course, formed and contribute significantly to the energy of binding as indicated by the greater values of that energy in netropsin and distamycin A than in SN 18071. The details of the results show, moreover, that the charged ends of the ligands are in the groove and do not exhibit direct interaction with the phosphates, a result confirmed by the X-ray analysis of reference 12.

The results presented in Table I refer to the DNA-ligand associations in vacuo. The fact that the free space computations account correctly for the observed specificity signifies that it is an <u>intrinsic property of these associations</u>. These primitive computations have been greatly refined since, in particular by the introduction of the solvent effect, either in the form of a distance dependent dielectric constant (22) or by an explicit introduction of water by a mixed discrete-continuum procedure (23). These refinements entirely confirm the essence of the previous results, while improving their numerical aspects. Thus e.g. they, naturally, decrease the absolute values of the interaction energies with respect to those in vacuo and bring the interaction energy of netropsin, the only one available experimentally (24,25), close to the observed value (exp. -11.2 kcal/mole, theor. -16.7 kcal/mole).

3) Parallel to these developments, a different suggestion as to the source of specificity in netropsin-DNA interactions was made by Dickerson and al. (26), who upon interpreting the crystal structure of the netropsin-dodecamer complex proposed that "the base specificity that makes netropsin bind preferentially to runs of AT base pairs is provided not by hydrogen bonding but by close van der Waals contacts between adenine C-2 hydrogens and CH groups on the pyrrole rings of the drug molecule". We shall explicit our view on this suggestion shortly in relation with the presentation and discussion of a particularly exciting line of investigation which consists of a search for GC minor groove specific ligands. This has developped in recent years into a large scale enterprise, with a number of successive stages.

III SEARCH FOR G-C specific minor groove binding ligands

A. Lexitropsins

In the last reference mentioned (26) Dickerson et al. suggested that modifying netropsin by replacing its pyrrole(s) by imidazole(s), the ring nitrogen of which may form a hydrogen bond with the NH_2 group of guanine, could yield analogues capable of recognizing preferentially GC base pairs. Such analogues were named "lexitropsins". Figure 1 indicates the chemical formulae and the denomination which we shall use for three representative derivatives of this class.

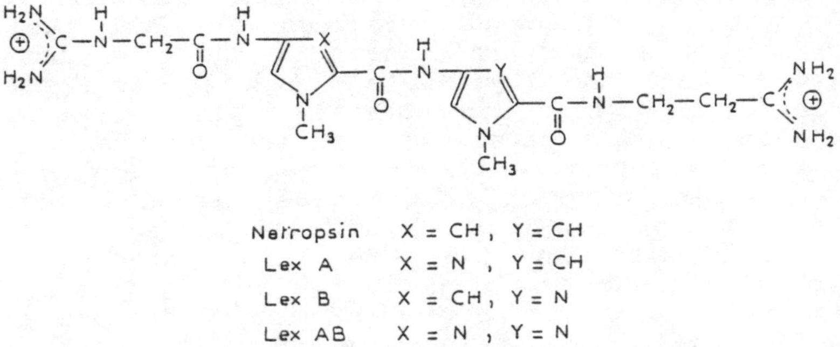

Figure 1 : Formulae and symbols for lexitropsins.

An experimental exploration of the binding and specificities of such lexitropsins (27,28) did not fulfill the main expectation of Dickerson's model. Thus although the presence of the imidazole groups in lexitropsins was found to permit binding to GC base pairs and although the progressive introduction of imidazoles resulted in a progressively decreasing preference for AT binding sites and a correspondingly-increasing acceptance of GC such sites, a preference for GC sites was not observed.

Table II

Complexation energies of netropsin and lexitropsins with poly(dA) . poly(dT) and poly(dG) . poly(dC) (kcal/mole).

Ligand	Poly(dA).poly(dT)	Poly(dG).poly(dC)	$\Delta_{complex}$ (AT)-(CG)
Netropsin	- 58.7	- 41.6	- 17.1
Lex A	- 55.0	- 40.3	- 14.7
Lex B	- 54.2	- 42.9	- 11.3
Lex AB	- 50.1	- 46.6	- 3.5

The theoretical exploration of this problem (29) provided an explanation of the observed situation. The essential results are summarized in Table II which shows that : 1) the complexation energy of the four ligands with poly(dA).poly(dT) decreases progressively with the substitution of one and then two pyrroles by imidazoles ; 2) the complexation energy of the ligands with poly(dG).poly(dC) increases progressively in the same circumstances ; 3) the preference for the AT sequence is conserved for all the ligands, although it decreases with the number of imidazoles incorporated in place of pyrroles.

Both the overall persistence of the preference for AT sequences and its progressive decrease upon the replacement of the pyrrole rings by imidazoles are in remarkable agreement with the experimental data. At the same time, the theoretical results invalidate the interpretation of the origin of AT specificity of netropsin and of distamycin A, as being due to contacts between adenine C2 protons and the pyrrole β protons. The same preference is still observed in the lexitropsins in which these contacts are abolished and in which, moreover, new hydrogen bonds are formed between the imidazole rings and guanines. Furthermore, details of the computations show that the dominance of the AT sequence over the GC sequence in all these interactions is due essentially to the greater value of the electrostatic component of the interaction energy with the former, reemphasizing thus the significance in this "specificity" of the stronger concentration of the molecular electrostatic potential in its minor groove. Obviously the new hydrogen bonds of lexitropsins are incapable of compensating completely for the initial electrostatic advantage of the minor groove of AT over that of GC sequences.

B. Isolexins

A second attempt to create improved candidates which would bind to the minor groove of GC sequences has as its starting point an observation by Goodsell and Dickerson (30) that the spread of the pyrrolecarboxamide repeat unit of netropsin is inherently too long for perfect matching with the DNA base pairs. A computer search led these authors to propose that analogs isohelical with DNA and thus ensuring a better binding could be obtained in the form of pyrrole-amine and pyrrole-ketone analogs of netropsin, in which the netropsin backbone is shortened by eliminating either the C=O or the N-H group of the amide units. The authors estimated then that by suitably placing pyrrole or furan rings in these sequence reading ligands (called isolexins), systems could be obtained capable to decipher the appropriate desirable DNA fragments : the pyrroles, following their previous idea (26) by van der Waals contacts between their N hydrogen and the C2 hydrogen of adenine, and the furans by a hydrogen bond between their oxygen and the NH_2 group of guanine.

It must be remarked immediately that the concept of H-H van der Waals contacts as factors of specificity, already disputable in the case of lexitropsins, is still less acceptable for the isolexins. The N-H group of their pyrrole, which is now the one oriented towards DNA, is obviously destined to act as an H-donor for hydrogen bond formation with the N3 atom of adenine or the O2 atom of thymine. An illustration of a refined scheme in which both heteroaromatic pentacycles function as H-bond forming units with the appropriate receptors on DNA, which will be used here, is given in Fig. 2 (31).

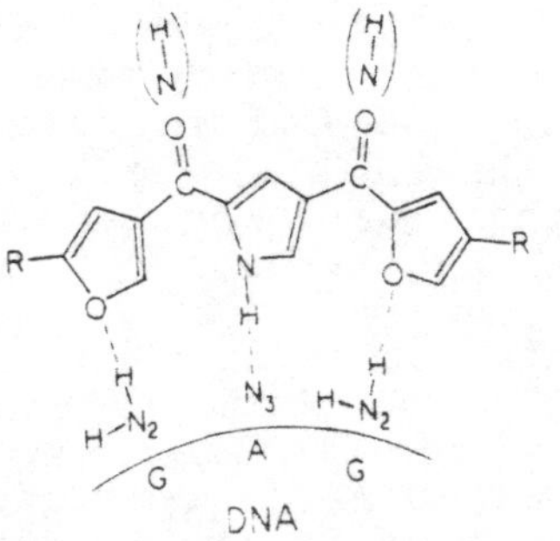

Figure 2 : Hydrogen bonding possibilities of the "furan-pyrrole-furan" isolexin.

It must be underlined that the isolexins not only produce a shortening of the ligand to dimensions isohelical with DNA but imply also a substantially modified scheme of interactions with this biopolymer : this scheme involves now only the heteroaromatic rings (and possibly the end groups) but no more the groups bridging the rings which, whether C=O or N-H, are directed towards the exterior of the complex.

The proposal of ref. 30 which considers the heteroaromatic pentacycles of isolexins as the only source of specificity suffers from two more drawbacks. In the first place, based on purely steric considerations it neglects the consideration of the electrostatic effects which, as we have seen, are essential in determining the intrinsic preference of this type of ligands for the minor groove of AT sequences. Further, in this case the possibility of occurence of a complementary perturbation, due to another electronic effect, is overlooked. Thus in the proposal of Goodsell and Dickerson the only important factor is the shortening of the amide linkage and no distinction is made whether this occurs by leaving the C=O or N-H group. The nature of these groups (to which we shall refer in the present context as linkers), oriented towards the outside and not taking thus any direct part in the binding with DNA, was not considered as having a possible significance for the variation of specificity.

That this could be but _a priori_ need not be the situation is evident if one considers the differences in the electronic properties of the C=O and N-H groups, in particular when they are engaged in conjugation with π electronic systems, as it is the case in isolexins. Quantum chemistry teaches us that while >C=O substracts π electrons from such systems, >N-H provides them with an excess of such electrons. It also indicates, that considering both the π and σ electrons, the C=O bond is associated with a dipole moment of about 2.4 D, with the overall direction C^+O^-, while the N-H bond is associated with a dipole moment of only about 1.3. D, with the overall direction N^-H^+. This situation must lead to significantly different dipole moments, both in magnitude and in direction, in isolexins utilizing the C=O or the N-H linkers. The natural question then arises : could these different electronic properties of the linking C=O or N-H groups of isolexins have any significant effect on the binding efficiencies and specificity in the interaction of these compounds with DNA ?

The answer to this question is provided in Table III (parts a and b) which presents the results of a theoretical exploration (31) of the binding specificities of the model isolexins V-VIII, composed of the furan-pyrrole-furan ring system which, when only the hydrogen-bonding possibilities are considered, appears appropriate for reading the base sequence GAG. These isolexins were considered with the two proposed "linkers" : C=O or N-H. DNA was represented by fragments 7-base pair long, one of which was a pure stretch of AT base pairs, the others containing a central GAG triplet. For reasons which will become clear shortly the terminals of the isolexins (R in Fig. 2) consist first of two cationic groups ($R{=}CH_2{-}CH_2{-}C(NH_2)^+$) which make them mimick the charged state of netropsin and of the previously studied lexitropsins, and in a second stage of neutral CH_3 groups.

In table III and in the subsequent one, δE^{DNA} represents DNA distorsion energy, δE^{Lig} ligand distorsion energy, $E^{DNA-Lig}$ the DNA-ligand interaction energy (composed of Lennard-Jones, electrostatic and polarization contributions) and E_C the total complexation energy.

The results of Table IIIa pertaining to the charged isolexins carrying two cationic terminals (V-VI) indicate that the best complexation energy for both ligands studied is obtained for the homogeneous sequence AAAAAAA despite the fact that their base reading moieties were designed to prefer the sequence GAG. The result is valid for both linkers, N-H and C=O, the differential effect of which is

Table IIIa

Complexation energies and their components for the interaction of dicationic isolexins (V, VI) with DNA fragments of different base sequences (values in kcal/mole).

DNA SEQUENCE	DNA δE	Lig δE	DNA - Lig E				E_C
			Lennard-Jones	Electro-static	Polari-zation	Total	
A. C=O group used as linker between base reading moieties (compound V)							
AAAAAAA	6.8	14.1	- 45.0	- 108.0	- 0.4	- 153.4	- 132.5
AAGAGAA	14.9	15.0	- 46.7	- 106.9	- 0.4	- 154.0	- 124.1
AACACAA	7.6	16.5	- 42.8	- 104.7	- 0.4	- 147.9	- 123.8
GGCACGG	6.0	8.1	- 44.5	- 86.9	- 0.4	- 131.8	- 117.7
B. NH group used as linker between base reading moieties (compound VI)							
AAAAAAA	7.2	16.3	- 44.3	- 94.6	- 0.4	- 139.3	- 115.8
AAGAGAA	14.0	17.3	- 45.2	- 94.7	- 0.4	- 140.3	- 109.0

Table IIIb

Complexation energies and their components for the interaction of neutral isolexins (VII, VIII) with the DNA fragments of different base sequences (values in kcal/mole).

DNA SEQUENCE	DNA δE	Lig δE	DNA - Lig E				E_C
			Lennard-Jones	Electro-static	Polari-zation	Total	
A. C=O group used as linker between base reading moieties , CH_3 terminals (compound VII)							
AAAAAAA	0.2	0.9	- 34.2	- 18.6	- 0.3	- 53.0	- 51.9
AAGAGAA	0.2	0.6	- 34.1	- 14.6	- 0.1	- 48.8	- 48.0
B. NH group used as linker between base reading moieties, CH_3 terminals (compound VIII)							
AAAAAAA	0.2	0.5	- 32.4	- 1.3	- 0.2	- 33.9	- 33.2
AAGAGAA	0.4	0.5	- 32.8	- 3.9	- 0.1	- 36.9	- 36.0
GGCACGG	1.0	0.5	- 31.9	- 7.1	- 0.2	- 39.2	- 37.7

noticed only in the smaller values of the complexation energy found for the isolexin with the N-H linkers and also in the smaller value of the separation between the complex formed with the pure A sequence and the most favourable one formed with the oligomer containing the cognate polynucleotide sequence namely AAGAGAA : the separation is 8.4 kcal/mole in the isolexin with C=O linkers and 6.8 kcal/mole in the isolexin with N-H linkers. Thus these dicationic ligands, just as the lexitropsins, do not fulfill the expectation of being able to serve as mixed base sequence specific ligands. In spite of their structural predestination to establish preferential H-bonds with the sequence GAG they still prefer to bind to the pure A strech. This result strikingly confirms the limitations of models of interaction specificity based uniquely on formal geometrical considerations and H-bonding possibilities. It also definitely eliminates the possibility for CH van der Waals contacts as playing a role in AT specificity of groove binding ligands. These isolexins simply lack such contacts.

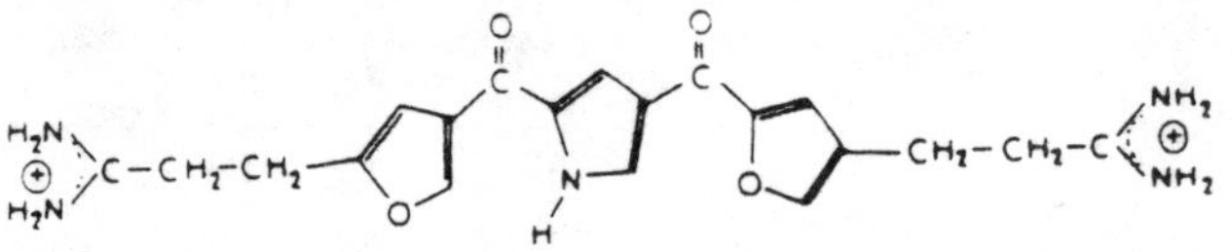

V

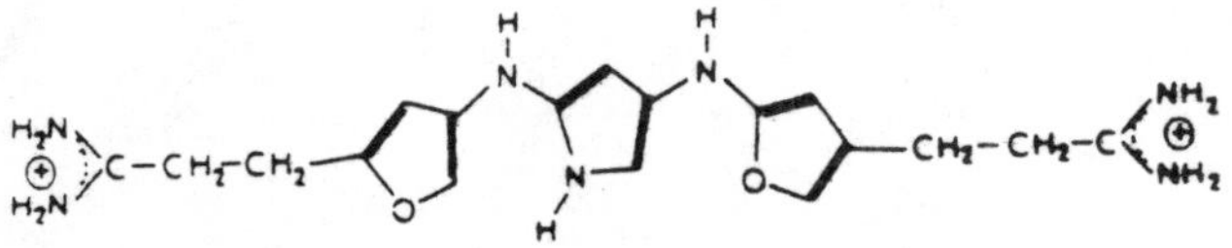

VI

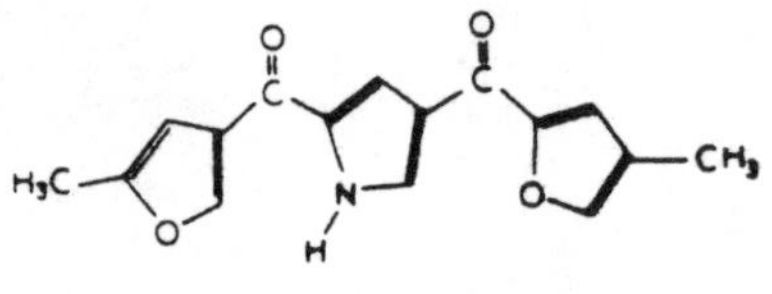

VII

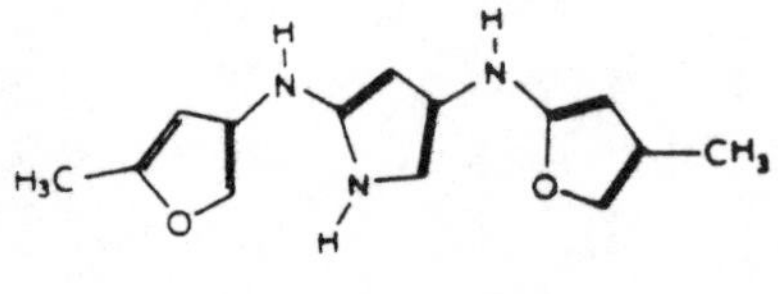

VIII

All the results obtained so far with dicationic groove binding ligands point thus to the large influence exercized upon their interaction with DNA by their charged ends, draining them preferentially towards the electrostatically privileged minor groove of AT sequences. It seemed therefore useful to investigate the effect of suppressing these charges. This was done by studying the two neutral analogue isolexins VII and VIII. The results of the computations are indicated in Table IIIb.

The first immediate observation is that in these cases the values of all the energy components are, understandably, reduced with respect to the charged isolexins. This reduction is the strongest for the interaction energy and comes mainly from that of the electrostatic contribution. In fact the Lennard-Jones component now becomes the main source of complex stabilization.

The most important result concerns, however, the sequence specificity of these neutral isolexins : thus while the neutral ligand with two C=O linkers continues to prefer binding to the pure AT sequence, the ligand with two N-H linkers shows a preference for the AAGAGAA stretch (of 2.8 kcal/mole) and even a slightly greater preference (of 4.5 kcal/mole) for the GGCACGG stretch.

This striking reversal of the classical situation (AT specificity) is the more impressive as, obviously, it is related to the nature of the linking groups, C=O or N-H, which oriented externally with respect to the oligonucleotide do not take part directly in the binding process.

Inspection of the results shows that this different behaviour of the two lexitropsins springs essentially from the different evolution of the relatively small but decisive electrostatic component of their DNA-ligand interaction energy. A more elaborate interpretation can be provided in terms of the interaction of the C=O and N-H dipoles with the field of the oligonucleotide acceptors (31).

One of the immediate consequences of the results presented in Table III was to lead, rather straightforwardly but for the first time in theoretical computations, to the conception of a possible GC minor groove specific ligand. Thus these results naturally suggest that such a ligand could be formed by the isolexin IX analogous to isolexin VIII (endowed with NH linkers) but possessing the regular heteroaromatic sequence furan-furan-furan enabling it to form three consecutive hydrogen bonds to three consecutive GC base pairs. That such should indeed be the situation is verified by the computations related in Table IV which indicate, indeed, the preferential binding (by 3.8 kcal/mole) of that isolexin to the 5'-AAGGGAA-3' sequence rather than to the homogenous 5'-AAAAAAA-3' one. The anchoring of the drug is ensured by the three expected

H-bonds, with length, respectively of 3.19, 1.94 and 2.16 Å (starting from the G at the 5′ side).

Table IV

Complexation energies and their components for the interaction of the neutral isolexin IX with two DNA fragments (values in kcal/mole).

DNA SEQUENCE	DNA δE	Lig δE	DNA - Lig E Lennard-Jones	Electro-static	Polari-zation	Total	E_C
AAAAAAA	1.0	0.2	- 31.1	2.3	- 0.2	- 29.0	- 27.8
AAGGGAA	1.6	0.2	- 29.9	- 3.5	- 0.1	- 33.4	- 31.6

One of the goals pursued in many laboratories of modeling a ligand capable of binding preferentially to the minor groove of GC sequences has thus been achieved. It was obvious, however, that whatever its merits isolexin IX had to be considered essentially as a <u>prototype</u>, susceptible of many improvements. From that point of view, the understanding of the roles played by the different components of the isolexins (heteroatomic rings, linkers, end groups) in stabilizing their potential interactions with DNA fragments authorized, however, the expectation of rapid, constructive advances. These expectations were indeed rapidly fulfilled in the form of a new group of ligands, the vinylexins.

C. Vinylexins

Within the "classical" conception of the origins of specificity the most straightforward way of increasing GC specificity of prototype IX, would the replacement of its furans by a better hydrogen bond acceptor. Indeed, a priori furan is not the best possible such acceptor, its oxygen atom having lost a part of its lone pair electrons towards the conjugated π-electron system of the ring. From this point of view the secondary nitrogen atom of imidazole, for example, should be a better acceptor as it preserves its lone pair in its entirety and even acquires an excess negative charge through conjugation effect.

Explicit computations (32) performed for the analog X of IX in which the furans have been replaced by methylimidazoles show however that this structural modification only moderately increases the GC selectivity in

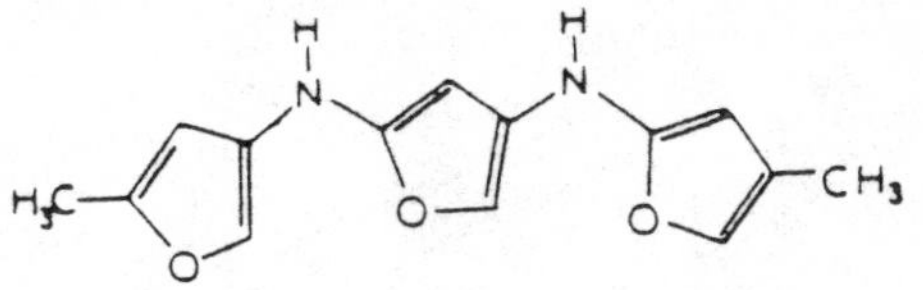

IX

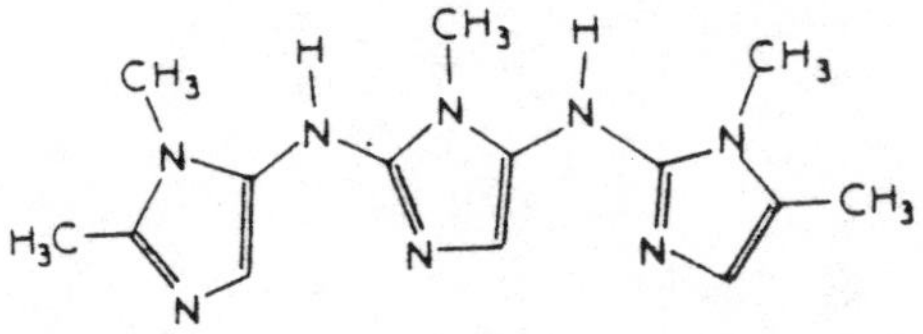

X

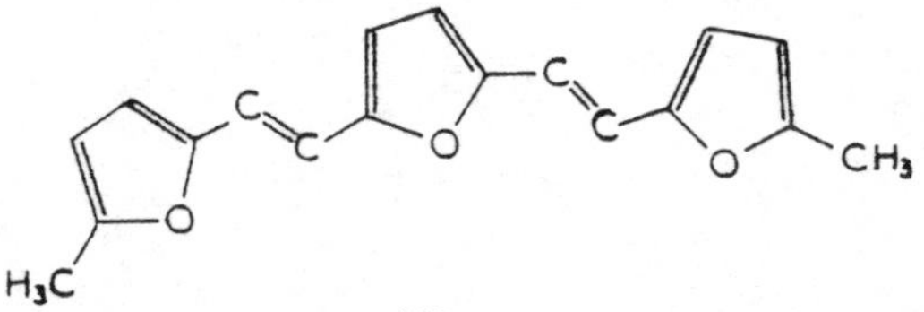

XI

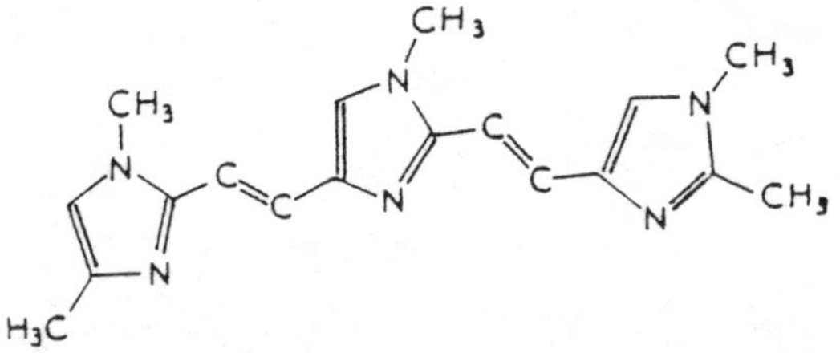

XII

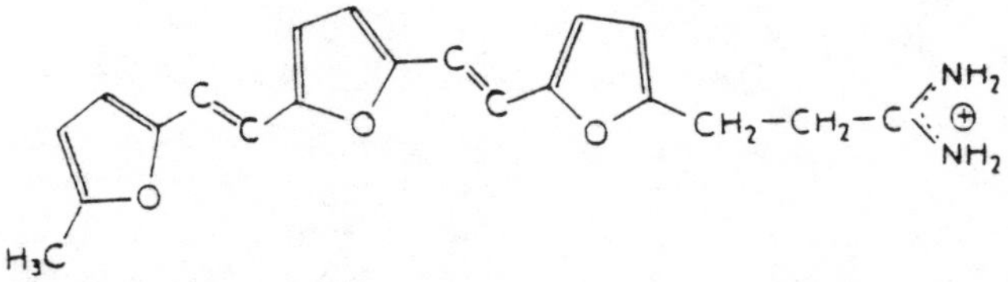

XIII

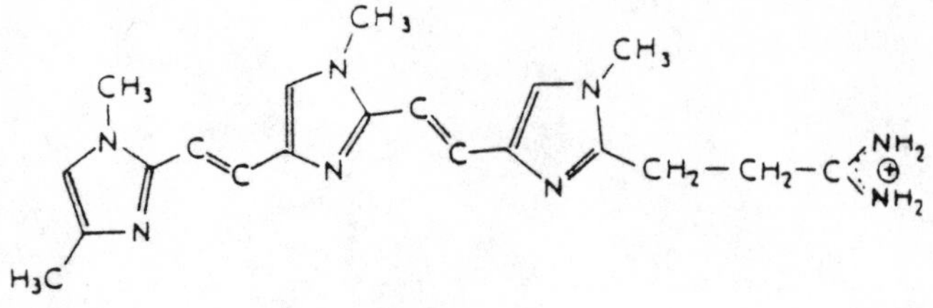

XIV

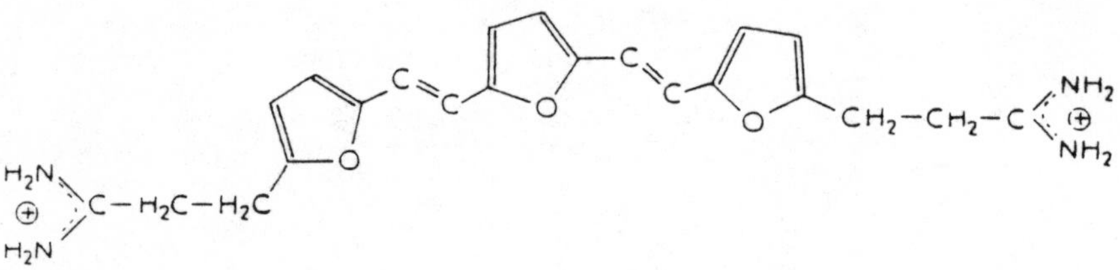

XV

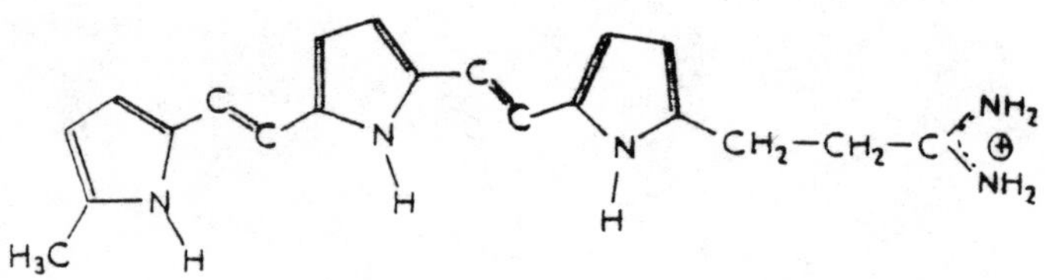

XVI

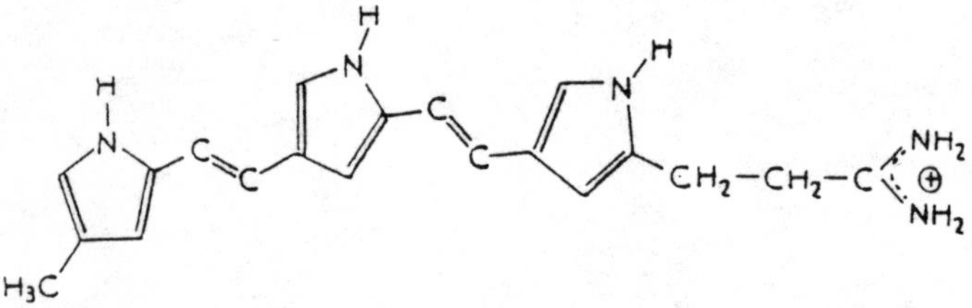

XVII

IX and moreover has no important effect on the complexation energy itself. The generation of an efficient GC minor groove specific agent by these means remains therefore problematic.

On the other hand, the results of ref. 31 and their analysis have brought into evidence the decisive importance of the nature of the "linkers" between the heteroaromatic rings for the production of GC minor groove specific ligands. The dipole moment of these groups is of particular significance in this respect, due to the effect of its interaction with the field of the receptor site within the nucleic acid. This situation prompted us to explore (32) the effect of a linker without dipole moment consisting of the ethylenic double bond. We shall call this type of compounds, represented by formulae XI and XII, vinylexins. The computations involved their interaction with a DNA receptor represented by an oligonucleotide double helix consisting of 7 nucleotide pairs. Two sequences were used AAAAAAA and AAGGGAA, referred to hereafter simply as AT and GC, respectively, in relation to the central 3 base pairs which constitute the ligand binding site.

Table V

DNA-ligand stabilization energy and specificity index of neutral vinylexins (kcal/mole).

LIG	SEQ	$\delta E(DNA)$	$\delta E(LIG)$	E(DNA-LIG)				E_C	δE_C
				L-J	ELEC	POL	TOT		
XI	GC	5.9	1.3	-38.7	-9.5	-0.2	-48.4	-41.2	- 7.7
	AT	5.6	0.9	-35.9	-3.8	-0.2	-40.0	-33.5	
XII	GC	7.4	1.8	-41.1	-16.3	-0.2	-57.6	-48.4	-14.0
	AT	6.7	1.7	-40.0	-2.6	-0.2	-42.8	-34.4	

The results of computations of the complexation energies of these ligands with these two representative AT and GC sequences are presented in Table V (31). The symbol δE_C, in the last column of this table, defined as the difference between the complexation energies E_C for GC sequence - AT sequence, may be considered as an index of specificity, negative values indicating GC specificity and positive values AT specificity. Our findings indicate that :

1) There is a substantial increase in the GC specificity of the three ligands, the E_C becoming -7.7 kcal/mole in XI and -14 kcal/mole in XII.

2) There is a concomitant increase in the complexation energy E_C itself, particularly for the interaction with the

GC receptor.

It is particularly remarkable that this doubly positive result was obtained by modifying fragments of the structure of the ligand (the linkers) which take no direct part in its interaction with DNA. This situation clearly demonstrates that, as stated in ref. 31, "specificity depends on all the steric and energetic components of complex formation". From the point of view of the numerical values of the interaction energy E(DNA-Lig) for the different compounds it is interesting to remark that in these neutral vinylexins the major part of their dominant contribution to the overall complexation energy E_C is the Lennard-Jones term. This component is also responsible in part for the GC specificity. The major contribution to specificity originates, however, from the electrostatic component, although its relative contribution to the interaction energy is significantly less than that of the Lennard-Jones term.

A possible further increase of the practical significance of these potential GC minor groove binding ligands may be achieved by introducing charged end groups. Thus it is known that a netropsin derivative in which the two cationic end groups of the natural antibiotic have been cut off continues to bind to poly(dA).poly(dT) (8). One may therefore expect binding to DNA to be also possible with the neutral vinylexins XI and XII. Nevertheless it is obvious that better binding may be expected with derivatives carrying cationic end groups. The problem is whether such cationic derivatives will still conserve GC specificity and, if so, to what extend. As shown explicitely in ref. 31 isolexins with both ends cationic always prefer AT sequences. Thus the dicationic analogue of IX looses the GC specificity for AT binding. The reason for this phenomenon is the overwhelming attraction of the doubly charged ligands by the elevated electrostatic molecular potential in the minor groove of AT sequences which overrides its possible GC preference due to its hydrogen bonding possibilities. An intermediate situation may be expected (33) for monocationic ligands and was explored with compounds XIII and XIV which are derivatives of the neutral vinylexins XI and XII in which one methyl end group has been replaced by a cationic CH_2-CH_2-$C(NH_2)_2^+$ group. The results obtained with these compounds are presented in table VI. They indicate :

1) A general substantial increase of the complexation energy of the ligands with DNA,

2) Only a moderate reduction of their GC specificity. XIII and XIV may therefore be recommended ligands for experimental investigations.

Table VI

DNA-ligand stabilization energy and specificity index of monocationic vinylexins (kcal/mole).

LIG	SEQ	δE(DNA)	δE(LIG)	E(DNA-LIG)				E_C	δE_C
				L-J	ELEC	POL	TOT		
XIII	GC	12.3	2.3	-43.0	-52.0	-0.3	-95.8	-81.2	- 8.0
	AT	9.2	2.2	-40.1	-44.2	-0.3	-84.6	-73.2	
XIV	GC	10.4	3.2	-47.3	-53.0	-0.3	-100.7	-87.1	- 9.3
	AT	8.0	2.5	-45.9	-42.2	-0.3	-88.3	-77.8	

Table VII

DNA-ligand stabilization energy and specificity index of bicationic furan vinylexin XV (kcal/mole).

LIG	SEQ	δE(DNA)	δE(LIG)	E(DNA-LIG)				E_C	δE_C
				L-J	ELEC	POL	TOT		
XV	GC	15.6	6.7	-50.1	-96.3	-0.4	-146.8	-124.5	- 3.5
	AT	9.5	8.2	-46.4	-91.9	-0.4	-138.7	-121.0	

We have also explored, for the example of vinylexin XV, the possible effect of two cationic end groups. The results presented in Table VII indicate, as could be expected, a further increase of the complexation energy but also a strong reduction of GC specificity, E_C now becoming reduced to -3.5 kcal/mole. The result obtained shows the importance of the non-dipolar nature of the vinylic linkers and at the same time supports our proposition that the optimal balance between the ligand complexation energy and GC sequence specificity will be obtained in the case of monocationic derivatives.

In relation to an earlier remark it is interesting to note that in the monocationic vinylexins the contribution of the electrostatic and Lennard-Jones components to the E(DNA-Lig) interaction energy are of the same order of magnitude. They both contribute to the GC specificity with again, generally, a larger contribution from the electrostatic term. On the other hand, the electrostatic component definitely predominates in the E(DNA-Lig) interaction energy of the dicationic vinylexin.

Monocationic vinylexins with furan and imidazole rings as hydrogen bond acceptors are thus as we have just seen, potentially good candidates for preferential binding to the minor groove of GC sequences. A question may then be raised as to whether similar vinylexins with, say, pyrroles as hydrogen bond donors will be able to function as ligands specific for the minor groove of AT sequences. If this is the case the C=C double bond would be confirmed as a generally appropriate linker, the orientation of the specificity of the ligand towards the minor groove of a given sequence depending then entirely on the nature of the heteroaromatic proton acceptor or donor rings.

Table VIII

DNA-ligand stabilization energy and specificity index of monocationic pyrrole vinylexins XVI and XVII (kcal/mole).

LIG	SEQ	δE(DNA)	δE(LIG)	E(DNA-LIG)				E_C	δE_C
				L-J	ELEC	POL	TOT		
XVI	GC	13.3	6.2	-37.3	-57.8	-0.2	-95.3	-75.8	29.7
	AT	10.8	3.2	-47.7	-70.3	-0.4	-118.5	-105.5	
XVII	GC	10.9	2.7	-39.0	-42.5	-0.2	-81.7	-68.1	23.3
	AT	9.7	1.5	-48.5	-53.8	-0.3	-102.6	-91.4	

Computations were consequently performed (32) for the monocationic vinylexin XVI containing pyrrole rings which are turned so as to be able to function as proton donors for H bond formation with the DNA receptor and also for compound XVII in which the orientation of the pyrrole rings is similar to that in netropsin or distamycin A, so that these moieties have no H-binding possibilities. The results presented in Table VIII indicate that both compounds should manifest remarkable AT specificity, somewhat greater in XVI than in XVII. This last result indicates that, as demonstrated by us previously in the case of netropsin and SN 18071, hydrogen bonds are not necessary for binding cationic ligands to the minor groove of AT sequences, although their presence does increase the complexation energy and AT specificity.

In the light of all these results it appears therefore that vinylexins are excellent candidates to compose a family of universal minor groove binding agents susceptible to bind to any chosen sequence as a function of the positioning of their proton donor and proton acceptor rings. Computations are under way to verify and extend this mixed scheme to longer base sequences.

This study thus confirms the insufficiency of considerations based solely on the notion of geometrical fitting, hydrogen bonding capabilities or van der Waals contacts for a correct estimation of GC versus AT specificities of groove binding ligands. Correct predictions can only be made by taking into consideration the overall electronic properties of the interacting species and explicitely calculating the energetics of DNA complex formation including all the relevant contributions (34).

Acknowledgment. The author expresses his deep thanks to the Association for International Cancer Research (Brunel and Saint-Andrews Universities, United Kingdom) for its support of this work.

References

1. P.B. Dervan : 1986, Science, p.464.
2. P.B. Dervan : 1987, in Molecular Mechanisms of Carcinogenesis and Antitumor Activity. B. Pullman and C. Chagas Eds. The Vatican Press, distributed by Adenine Press, New York, p.365.
3. P.K. Werner and P.A. Kollman : 1981, J. Comput. Chem. 2, p.287.
4. N. Gresh, P. Claverie and A. Pullman : 1984, Theoret. Chim. Acta 66, p.1.
5. N. Gresh, P. Claverie and A. Pullman : 1986, Int. J. Quantum Chem. 29, p.101.
6. R. Lavery : 1988, in Unusual DNA Structures, eds S. Harvey and R. Wells, Springer-Verlag Pergamon Press, New-York, p.189.
7. R. Lavery : 1988, in DNA Binding and Curvature eds W. Olson, M. Sundaraligam, M. Sarma and R.H. Sarma, Adenine Press, New York, p.191.
8. Ch. Zimmer and U. Wahnert : 1986, Progress Biophys. Molec. Biol. 47, p.31.
9. G.V. Gursky, V.G. Tumanyan, A.S. Zasedatelev, A.L. Zhuze, S.L. Grokhovsky and B.P. Gottikh : Acid-Protein Recognition, Ed. H.J. Vogel, Academic Press, p.189.
10. A.J. Krylov, S.L. Grokhovsky, A.S. Zasedatelev, A.L. Shuze, G.V. Gursky and B.P. Gottikh : 1979, Nucl. Acids Res. 6, p.289.
11. H.M. Berman, S. Neidle, Ch. Zimmer and H. Thrum : 1979, Biochem. Biophys. Acta 561, p.124.
12. M.L. Kopka, Ch. Yoon, D. Goodsell, P. Pjura and R. Dickerson : 1985, J. Mol. Biol. 183, p.553.
13. D.J. Patel and L. Shapiro : 1986, Biopolymers 25, p.707.
14. R.E. Klevit, D.E. Wemmer and B.R. Reid : 1986, Biochemistry 25, p.3296.
15. A.W. Braithwaite and B.C. Baguley : 1980, Biochemistry 19, p.1101.

16. B.C. Baguley : 1982, Molecular and Cellular Biochemistry 43, p.167.
17. B. Pullman and A. Pullman : 1981, Studia Biophysica 86, p.95.
18. A. Pullman and B. Pullman : 1981, Quarter Rev. Biophysics 14, p.289.
19. R. Lavery and B. Pullman : 1985, J. Biomol. Struct. Dyn. 2, p.1021.
20. A.V. Fratini, M.L. Kopka, H.R. Drew and R.E. Dickerson : 1982, J. Biol. Chem. 257, p.14686.
21. K. Zakrzewska, R. Lavery and B. Pullman : 1983, Nucl. Acids Res. 11, p.8825.
22. R. Lavery, K. Zakrzewska and B. Pullman : 1986, J. Biomol. Struct. Dyn 3, p.1155.
23. K. Zakrzewska, R. Lavery and B. Pullman : 1984, Nucl. Acids Res. 12, p.6559.
24. L.A. Marky, K.S. Blumenfeld and K.J. Breslauer : 1983, Nucl. Acids Res. 11, p.2857.
25. L.A. Marky, J. Curry and K.J. Breslauer : in Molecular Basis of Cancer, R. Rein Ed, Han R. Liss Inc. N.Y. Part B, p.155.
26. M.L. Kopka, Ch. Yoon, D. Goodsell, P. Pjura and R.E. Dickerson : 1985, Proc. Natl. Acad. Sci. USA 82, p.1376.
27. J.W. Lown, K. Krowicki, U.G. Bhat, A. Skorobogaty, B. Ward and J.C. Dabrowiak : 1986, Biochemistry 25, p.7408.
28. Ch. Zimmer, G. Luck, G. Burckhardt, K. Krowicki and J.W. Lown : 1987, in Molecular Mechanism in Carcinogenic and Antitumor Activity Eds. B. Pullman and C. Chagas, The Vatican Press, distributed by Adenine Press, New York, p. 339.
29. K. Zakrzewska, R. Lavery and B. Pullman : 1987, J. Biomol. Struct. Dyn. 4, p.833.
30. D. Goodsell and R.E. Dickerson : 1986, J. Med. Chem. 29, p.727.
31. K. Zakrzewska and B. Pullman : 1988, J. Biomol. Struct. Dyn. 5, p.1043.
32. K. Zakrzewska, M. Randrianarivelo and B. Pullman : J. Biomol. Struct. Dyn., in press.
33. K. Kissinger, K. Krowicki, J.C. Dabrowiak and J.W. Lown : 1987, Biochemistry 26, p.5590.
34. B. Pullman : Advances in Drug Research, in press.

THE ORIGIN, DEVELOPMENT AND SIGNIFICANCE OF THE HEITLER-LONDON APPROACH

Włodzimierz Kołos
Quantum Chemistry Laboratory
Department of Chemistry
University of Warsaw, Pasteura 1
02-093 Warsaw, Poland

ABSTRACT. The pioneering work by Heitler and London and its subsequent development by Slater and Pauling are recalled. Modern formulation of the classical valence bond (VB) theory is briefly sketched. Its fundamental contribution to our understanding of chemical bonds, of electronic and geometric structure of molecules and of their properties is reminded. Extensions of the VB method are indicated, and its particular advantages for treating long-range interactions are discussed.

1. INTRODUCTION

The concept of chemical structure was introduced around the middle of the XIX century. It had a huge impact on the development of chemistry, especially after 1874 when J.H. van't Hoff and J.A. le Bel postulated independently the tetrahedral arrangement of bonds around the carbon atom. The nineteenth century bonds, proposed when even the existence of atoms was questioned by many, had obviously no physical meaning. They were helpful in ordering numerous chemical facts but their nature was completely unknown.

The electronic theory of valence began in 1916 when G. N. Lewis [1] introduced the fruitful idea of sharing of two electrons by two atoms and thus forming, what is now called, the covalent bond. It rationalized a significant part of accumulated chemical evidence, however, it still could not explain neither the origin of bonding nor the nature of the interactions involved. Lewis' paper appeared three years after Bohr's formulation of his theory of the hydrogen atom and no more progress in understanding the most fundamental chemical facts could be made using the planetary model of atoms. Bohr's theory could not explain the stability of molecules, and the theory of Lewis, related to it, could not explain the postulated privileged position of an electron

J. Jortner and B. Pullman (eds.), Perspectives in Quantum Chemistry, 145–159.

pair.

A decade later quantum mechanics was developed and the new formalism was immediately applied to explain the electronic structure of molecules and to initiate a new branch of science that 10 years later was named [2] quantum chemistry.

It is difficult to connect the birth of quantum chemistry with a single event. On 27 January 1926 Erwin Schrödinger [3], working at the Zürich University, submitted to Annalen der Physik his paper 'Quantisierung als Eigenwertproblem', and this was the birth of the Schrödinger equation.

Soon after Schrödinger's work Werner Heisenberg [4], while staying in Copenhagen, published a series of papers in which he presented a quantum mechanical approach to many-particle systems. When considering two-electron atomic systems he showed that the correct lowest order wavefunction describing the interacting electrons was

$$\varphi_i(1)\varphi_j(2) \pm \varphi_i(2)\varphi_j(1) \tag{1}$$

where φ_i and φ_j are wavefunctions of the two electrons in the static spherical field of the nucleus, respectively. This form of the wavefunction had several important consequences. It is obvious, when we look at it, that it inspired Walter Heitler and Fritz London, who similarly as Schrödinger were working at that time in Zürich, to develop their theory of covalent chemical bonds. Their pioneering work [5] that we celebrate today was received by the editor of the Zeitschrift für Physik on 30 June 1927. An account of it had already been given at a meeting of physicists in Freiburg on 12 June 1927. Thus this date is regarded by many as the date of birth of quantum chemistry. Shortly thereafter, on 25 August 1927, Max Born and Robert Oppenheimer [6] submitted to Annalen der Physik their manuscript entitled 'Zur Quantentheorie der Molekeln' in which the authors performed separation of the electronic and nuclear motion in molecules (assumed without proof by Heitler and London) thus setting foundation for the theory of molecular spectra and for computation of potential energy surfaces. The first preliminary discussion of the electronic and nuclear motion in a molecule was given, however, by Edward U. Condon [7] in a paper submitted on 19 March 1927 to the Proceedings of the National Academy of Sciences. It contains a quantum mechanical justification of the mechanism of the electronic transitions in molecules, well known today as the Franck-Condon principle.

A historical interlude, however, seems to be appropriate at this place. Namely a paper by Friedrich Hund [8] is sometimes cited as the first publication on molecular orbital (MO) theory. This article, entitled 'Zur Deutung der Molekelspektren. I', and written when Hund, similarly as Heisenberg, was staying in Copenhagen, was submitted to the Zeitschrift für Physik on 19 November 1926, i.e., earlier than the Heitler and London paper. It contained, *inter alia*, correlation diagrams for one- and two-electron molecules, but from his purely qualitative considerations the author could not show whether the ground state potential energy curve of the hydrogen molecule had a minimum. Thus in 1926 the stability of the hydrogen molecule was still not understood. The first successful calculation of the electronic energy of a diatomic molecule was that of the Danish astronomer Øyvind Burrau [9] on the one-electron hydrogen molecular ion. In elliptic coordinates he performed separation of the variables and solved the resulting two ordinary differential equations. The computed binding energy of H_2^+ was in a reasonable agreement with the value obtained indirectly from experimental data. Burrau's paper entitled 'Berechnung des Energiewertes des Wasserstoffmolekel-Ions (H_2^+) im Normalzustand' was presented by N. Bohr to the Royal Danish Academy of Sciences and Letters on 17 December 1926. At a meeting of the Academy on 11 February 1927 it was accepted for publication, and appeared in print on 19 March 1927 (i.e. 5 weeks later!). Burrau's manuscript apparently had been known to E.U. Condon, who in 1927 had a postdoctoral position with Arnold Sommerfeld in Munich. On 18 April 1927 he presented a paper [10] to the Proceedings of the National Academy of Sciences and the results contained in this paper he reported [11] also at a meeting of the Physical Society of Berlin on 13 May 1927. The title of the paper was 'Wave Mechanics and the Normal State of the Hydrogen Molecule'. Condon quotes Burrau's [9] results for the hydrogen molecular ion and writes: "Turning now to the neutral (H_2) molecule one expects on the Pauli principle of assigning quantum numbers, that the two electrons will be in equivalent orbits". Thus this work is sometimes regarded as the first molecular orbital treatment of a molecule. Condon utilized Burrau's results and instead of computing the electron-electron interaction integral he scaled the corresponding contribution evaluated somewhat earlier by Albrecht Unsold [12] for the helium atom. In this way he obtained a reasonable potential energy curve for the hydrogen molecule. Thus, in conclusion, he felt entitled to emphasize that "Burrau's calculation of H_2^+ and the extension here to H_2 constitute the first quantum-theoretic quantitative discussion of the binding of atoms into molecules by electrons - the valence forces of chemistry". Please note that Condon's result was made public about two months earlier than that of Heitler and London. The early molecular orbital type approach, however, was directed mainly towards interpretation of molecular spectra. In studies of molecular structure and of chemical bonds it was overshadowed by the rapidly developing valence bond theory that originated from the Heitler and London paper.

Every chemistry student knows that the Heitler-London wavefunction for the hydrogen molecule has the form

$$\Psi = 2^{-1/2}[\chi_a(1)\chi_b(2) + \chi_a(2)\chi_b(1)] \tag{2}$$

where χ_a and χ_b are the hydrogen 1*s* orbitals centered around nuclei *a* and *b*, respectively. To simplify the notation the normalization factor resulting from nonorthogonality of the orbitals has not been included in Eq.(2). The singlet state spin function for a two-electron system is

$$\Theta = 2^{-1/2}[\alpha(1)\beta(2) - \alpha(2)\beta(1)] \tag{3}$$

Hence the total wavefunction for the two electrons in the hydrogen molecule can be expressed in the form

$$\Phi(1,2) = 2^{1/2}\mathcal{A}[\chi_a(1)\chi_b(2)\Theta] \tag{4}$$

where $\mathcal{A}$ is the antisymmetrizer

$$\mathcal{A} = \frac{1}{N!}\sum_P(-1)^P P \tag{5}$$

N denotes the number of electrons, and P the permutation operators.

It was a great achievement of Heitler and London when, using wavefunction (2), they obtained a bound hydrogen molecule. However the actual value of the binding energy was still out of reach. Special work and another publication were needed to evaluate the exchange integral. It was carried out by Yoshikatsu Sugiura [13] who, working in Göttingen on atomic collisions, thanks to M. Born, became acquainted with the manuscript of the Heitler and London work. Already on 30 August 1927 he submitted his paper with the first complete and truly *ab initio* treatment of the hydrogen molecule. The value of the binding energy and of the equilibrium internuclear distance calculated in this way were in a reasonable agreement with experiment. In subsequent years physical intuition was utilized to improve the wavefunction (2) proposed by Heitler and London. In 1928 S.C. Wang [14] performed the exponent optimization in the atomic orbitals χ. In 1931 N. Rosen [15] allowed for deformation of the 1*s* orbitals and in 1933 S. Weinbaum [16] included ionic terms.

Finally a deeper understanding of the approach initiated by Heitler and London, and of its relation to the molecular orbital method was provided in 1949 by Charles A. Coulson and Inga Fischer [17].

Extension to more complex molecules, diatomic and polyatomic did not wait, however, until the hydrogen molecule was fully understood. Already in 1928 Linus Pauling [18] published a paper on shared electron-pair theory of the chemical bond. Three years later, due to the independent work by John C. Slater [19] and Linus Pauling [20] the Heitler-London approach was generalized to arbitrary polyatomic molecules. Under the name of Heitler-London-Slater-Pauling (HLSP) or valence bond (VB) method it expanded rapidly and seemed to offer at least qualitative answers to any chemical problem. This stage of development of the VB theory culuminated in 1940 in The Nature of the Chemical Bond by L. Pauling [21].

For several years both Heitler and London continued working on molecular problems. Heitler [22] developed a still different formulation of the VB theory, more suitable for larger distances between the interacting atoms. London considered explicitly large interatomic separations and starting from the Heitler-London wavefunction for the hydrogen molecule he laid foundation for the quantum mechanical theory of intermolecular interactions [23, 24]. We shall come back to these extensions in Section 4.

2. HLSP OR VB THEORY

Following Gerratt [25] a general formulation of the VB theory will now be briefly sketched. Let us denote by

$$\Psi_{Sp} = \Psi_{Sp}(1, 2, \ldots N) \tag{6}$$

a set of orthonormal N-electron spatial wavefunctions, that are degenerate eigenfunctions of the molecular Hamiltonian, and by

$$\Theta^{N}_{SMp} = \Theta^{N}_{SMp}(\sigma_1, \sigma_2, \ldots \sigma_N) \tag{7}$$

a set of orthonormal N-electron spin functions that are eigenfunctions of the total spin angular momentum and of its z component, with S and M denoting the corresponding spin quantum numbers, respectively. Then, as shown by Wigner [26], the total wavefunction that, in addition to being an eigenfunction of S^2 and S_z, is antisymmetric with respect to electron permutation, must have the form

$$\Phi_{SM} = (f_S^N)^{-1/2} \sum_{p=1}^{f_S^N} \Psi_{Sp} \Theta^N_{SMp} \tag{8}$$

where

$$f_S^N = \frac{(2S+1)N!}{(N/2+S+1)!(N/2-S)!} \tag{9}$$

An acceptable wavefunction can also be constructed from an arbitrary spatial N-electron wavefunction $\Psi(1, 2, \ldots N)$. If we denote (cf. Eq.(4))

$$\Phi_{SMp} = (N!)^{1/2} \mathcal{A}(\Psi\Theta^N_{SMp}), \tag{10}$$

for $p = 1, 2, \ldots f_S^N$, then the most general total wavefunction Φ_{SM} resulting from Ψ can be expressed as

$$\Phi_{SM} = \sum_{p=1}^{f_S^N} c_p \Phi_{SMp} \tag{11}$$

The antisymmetrizer ensures the proper antisymmetry of each Φ_{SMp}, and hence of Φ_{SM}, and the summation over p includes all possible coupling schemes of electron spins (see below) resulting in the total spin S. The approximate value of the energy and the corresponding values of the expansion coefficients can be calculated variationally. Since only the $S = M = 0$ states will be considered we can simplify the notation by deleting the indices S and M.

Let us now consider a system of two atoms A and B, and let us represent their spatial wavefunctions as

$$\Psi_A = \psi_{A1}(1)\psi_{A1}(2)\psi_{A2}(3)\psi_{A2}(4)\ldots\times$$

$$\times\ \psi_k(2k)\psi_{A,k+1}(2k+1)\ldots\psi_{A,K-k}(K) \tag{12}$$

$$\Psi_{B} = \psi_{B1}(1)\psi_{B1}(2)\psi_{B2}(3)_{B2}(4)\ldots\times$$

$$\times\, \psi_{l}(2l)\psi_{B,l+1}(2l+1)\ldots\psi_{B,L-l}(L) \qquad (13)$$

respectively. The atoms A and B are assumed to have k or l electron pairs, and $K - 2k$ or $L - 2l$ single electrons, respectively.

According to Eq.(4) the wavefunction for the composite system AB is

$$\Phi = \sum_{p=1} c_{p}(N!)^{1/2}\mathcal{A}[\psi_{A1}\psi_{A1}\psi_{B1}\psi_{B1}\psi_{B1}\psi_{A2}\psi_{A2}\ldots\times$$

$$\times\, \psi_{A,k+1}\psi_{B,l+1}\psi_{A,k+2}\psi_{B,l+2}\ldots\ \psi_{A,K-k}\psi_{B,L-l}\Theta^{N}_{p}] \qquad (14)$$

where $N = K + L$ is the total number of electrons, the numbers of single electrons in A and B are assumed to be equal, and the index p specifies a particular pairing of the spins in Θ^{N}_{p}. There are several possible ways to determine the spin functions Θ^{N}_{p}, as described in details by R. Pauncz [27]. A common procedure consists in building up the spin function by coupling successively the spins of the individual electrons. The branching diagram [28] can be used to find all the coupling modes leading to $S = 0$.

Another possibility consists in first coupling electron pairs into singlets and using these pair functions to form complete $S = 0$ spin functions for the system. One can easily verify that for even N the above mentioned procedure generates $N![2^{N/2}(N/2)!]$ complete singlet spin functions. For $N > 2$ this is larger than f^{N} and therefore the spin functions constructed in this way are not linearly independent. The Rumer [29] method can be used to get a linearly independent set.

The ordering of orbitals in Eq.(14) is immaterial if the most general linear combination of spin functions is used. One particular ordering, however, is privileged. Let us consider, e.g., the nitrogen molecule [30, 25]. In this case one of the terms in the sum over p in Eq.(14) is

$$\Phi_p = (14)!^{1/2}\mathcal{A}[1s_A 1s_A 1s_B 1s_B 2s_A 2s_A 2s_B 2s_B \times$$

$$\times 2p_{xA} 2p_{xB} 2p_{yA} 2p_{yB} 2p_{zA} 2p_{zB} \Theta_p^{14}] \tag{15}$$

If each consecutive pair of electrons in Eq.(15) is coupled into a singlet one gets a "perfect pairing" function which gives the largest contribution in Eq.(14). It describes two inner pairs of electrons, two $2s$ lone pairs and a triple bond in N_2.

Let us now consider a system of N atoms, each having one valence electron. The remaining electrons can be disregarded and the wavefunction for the valence electrons can be expressed as

$$\Phi_p = (N!)^{1/2}\mathcal{A}[\psi_1\psi_2 \ldots \psi_N\Theta_p] \tag{16}$$

For $N = 6$, and ψ_J denoting $2p\pi$ orbitals of carbon atoms, Eq.(16) can represent a π-electron wavefunction of benzene. If the molecule under consideration has some symmetry one should use linear combinations of Φ_p which form bases of irreducible representations of the proper symmetry group. The coefficients in these combinations are determined by standard methods of group theory.

If the spin functions Θ_p in Eq.(16) are formed only from singlet pairs, and the Rumer method is used to select the linearly independent set, one gets the classic valence bond functions. The coupled electron pairs in Eq.(16) are associated with specific pairs of atoms and can be considered as representing bonds between these atoms. Each wavefunction obtained in the above way represents a certain distribution of bonds in the molecule and is called a covalent structure. For instance, in the case of the six π electrons in benzene one gets the well known Kekulé and Dewar structures.

The total wavefunction of the system represents a superposition of these limiting structures, and is expressed as a linear combination of the corresponding functions (16) with coefficients determined by the variational method. If there are two identical orbitals in the wavefunction (16) it corresponds to an electron transfer and thus represents an ionic structure. With two such pairs one gets a doubly ionic structure etc.

All this sounds almost trivial in 1988 but it was different more than half a century ago when the VB theory was being developed and applied to numerous problems. It gave quantum mechanical meaning to the strokes or double dots representing bonds in classical formulas, and rationalized various molecular properties. The clear chemical meaning of the VB wavefunctions, their direct relation to the accepted Lewis' theory, made the method very appealing for chemists.

The VB method of describing the electronic structure of molecules was found to be advantageous not only for treating chemical bonds but also in various spectroscopic problems. It is, e.g., very suitable for describing states for which the wavefunction changes its character with internuclear separation. For instance, a maximum in the $A\ ^1\Pi$ state potential curve of BH has been interpreted in an elegant way [31] as arising from avoided crossing of two curves obtained from wavefunctions constructed from the $1s$ wavefunction of the H atom and from $2s^2 2p$ or $2s2p^2$ wavefunctions of the B atom, respectively.

The concept of VB structures is also useful for interpreting complex wavefunctions obtained in more advanced approaches. If a very accurate wavefunction is computed it is usually impossible to give it a clear physical interpretation. This, however, can be achieved by decomposing the wavefunction in terms of VB structures. Such a decomposition has been carried out, e.g., for the $B\ ^1\Sigma_u^+$ state of H_2 [32] providing a clear interpretation of the accurate wavefunction and of the type of interactions (covalent, $1s2s$, $1s2p\sigma$ and ionic) involved at various internuclear separations.

One can hardly exaggerate the important role that the VB theory has played in the development of quantum chemistry, and of the theory of molecular structure in particular. Its qualitative offspring, developed by Pauling, under the name of the resonance theory, gained wide acceptance. It gave a consistent explanation of a huge range of chemical phenomena, e.g., the covalent and ionic character of bonds, bond strength, saturation of valence, multiple bonds and their properties, geometry of molecules, properties of coordination ions, of conjugated and aromatic molecules etc. At present the concept of hybrid orbitals, of hybrid molecular structures do not belong to the most favorite ones in modern quantum chemistry, but they are still in everyday use among inorganic and organic chemists when discussing molecular binding and molecular structure.

3. EXTENSIONS OF THE VB APPROACH

From the very beginning, i.e., from the Heitler and London paper, it was known that overlap was essential for binding. On the other hand nonorthogonality of orbitals was the source of very serious numerical difficulties in computation of matrix elements. Attempts were made to remove these difficulties by orthogonalizing the orbitals belonging to different atoms [33, 34]. If, however, orthogonalized orbitals are used in the original Heitler-London wavefunction for the hydrogen molecule one does not get any minimum of the energy as function of the internuclear distance, and no stable molecule [35]. Configuration interaction is needed to get binding, but this brings back all the numerical difficulties.

The undisputable advantages of the VB method, however, stimulated work on making it practically applicable, and considerable progress has been made in dealing with nonorthogonal orbitals [36]. A natural extension of these efforts was directed towards increasing the accuracy of the results. A possible way of improving the VB wavefunction was indicated a long time ago by C.A.Coulson and I.Fischer [17] who elaborated an old idea of J.C.Slater [37].

In the early years of quantum chemistry the MO and VB methods were developed independently, and there was no clear opinion on their respective advantages and disadvantages. The MO description of molecules is attractive because it follows the same lines as that of atoms in the one-electron approximation, but it is well known that it fails completely for large internuclear separations between interacting open-shell systems. In contrast, the VB approach describes correctly the dissociation of molecules and therefore it is much more suitable than the MO method for calculation of potential energy surfaces.

In 1949 Coulson and Fischer [17] have shown, however, that for the hydrogen molecule both methods, if slightly generalized, become equivavalent. This happens when the Heitler-London wavefunction is supplemented with ionic terms, whereas in the MO approach some CI is included by adding a second configuration with both electrons described by the lowest antibonding orbital. This is well known today. In Poland a chemistry student who does not know it will not pass the obligatory quantum chemistry examination in the second year of his studies. It has been shown, however, by Coulson and Fischer that the same wavefunction can be transformed to a still different form which looks exactly like the Heitler-London wavefunction (2) but the orbitals are represented as

$$\chi_a = 1s_a + k1s_b, \qquad \chi_b = 1s_b + k1s_a \tag{17}$$

This makes it a very interesting wavefunction. For $k = 0$ it is identical with the original Heitler-London wavefunction, whereas for $k = 1$ it changes into the standard molecular orbital form. If the parameter k has an intermediate value it corresponds to inclusion of ionic terms in the VB treatment or of CI in the MO approach. On the other hand, however, the functions represented by Eq.(17) can be regarded as deformed orbitals of atom A and B, respectively. Thus a suitable deformation of the orbitals in the VB method can give an analogous effect as CI in the MO method.

An important generalization of the VB theory made in recent years by Goddard with coworkers [38], and Gerratt with coworkers [39,40] included deformation of orbitals. However, lack of time does not permit me to go into the details of these and other recent developments of the VB theory.

To conclude this section let me only briefly mention an interesting approach developed in 1953 by Hurley, Lennard-Jones and Pople [41]. Recognizing the importance of nonorthogonality they preserved it within bonds but imposed orthogonality constraints on orbitals associated with different bonds. This formulation, sometimes called Extended Valence Bond (EVB) method, is limited to systems with well localized two-electron bonds, and no (or only weak) interaction between them. For such systems, however, it gives a quantum mechanical justification of two well established rules derived from chemical evidence and concerning the binding energies of polyatomic molecules. It shows that the total binding energy can be expressed as a sum of bond energies, and that the energy of a given type of bond is approximately the same, independently of its surrounding, thus justifying the transferability of bond energies. In the limiting case, when two orbitals in each bond are identical, and are of two-center type, the EVB method becomes related to the MO theory. If, however, the identity constraint is relaxed one gets two different orbitals localized mostly on one or another atom involved in the bond, respectively. In this special case the EVB method is related to the treatment of the hydrogen molecule by Coulson and Fischer [17], and of the hydrogen and hydrogen fluride molecules by Mueller and Eyring [42]. It is also related to the very general spin-coupled VB method successfully developed more recently by Gerratt and Raimondi [39]. Today, however, the EVB theory is only of historical interest. It was originally developed to overcome the difficulties caused by the nonorthogonality of the orbitals used. As already mentioned, in later years many approaches have been proposed to deal with this difficulty, and they have been successfully applied to solve a variety of chemical problems.

4. LARGE INTERATOMIC SEPARATIONS

We have already stated that in contrast to the MO approach the VB method is suitable for computing energies for large distances between the interacting open-shell atoms. However, the above statement is true only if a better approximation than the "perfect pairing" is used. This is caused by the fact that the "perfect pairing" wavefunction does not dissociate into isolated atoms in well defined states but rather in the so called "valence states" which are mixtures of physical *L-S* coupled states.

The above mentioned difficulties disappear if the total wavefunction in the form of the expansion (14) is used rather than the single term corresponding to the "perfect pairing". One can also proceed in a different way, viz., starting from atoms in well defined *L-S* states, as recommended by Heitler [22]. For instance, in the case of the nitrogen molecule one can construct for each N atom a wavefunction describing the ^4S ground state for the p^3 configuration. These two wavefunctions can next be coupled to yield the $^1\Sigma_g^+$ ground state of the nitrogen molecule. Obviously, a wavefunction of this type can give a proper description of dissociation into the ^4S nitrogen atoms.

The correct treatment of molecular dissociation makes the VB method particularly suitable for computation of potential surfaces. This is in contrast to the MO method which, without configuration interaction, cannot give reliable interaction energies for large separations between interacting open-shell systems. The VB *ansatz* represents therefore the most natural zero-order wavefunction in any perturbation type theory of molecular interactions. Pioneering work in this direction was carried out by London [23,24] who first derived, in the second order of the perturbation theory, the contribution to the interaction energy which is due to electron correlation, and called it dispersion energy. His approach was taken up over 30 years later by many scientists who developed various formulations of the so called symmetry adapted perturbation theory of molecular interactions [43]. If SCF wavefunctions are used for the interacting subunits the first order energy in these formalisms represents the energy of interaction of unperturbed Hartree Fock subsystems, whereas the second and higher order energy contributions account for all the deformation and correlation effects.

It should be also pointed out that in their pioneering paper [5] that we celebrate today Heitler and London explicitly considered not only the chemical H-H bond but also interactions between closed shell systems. For two hydrogen atoms they have shown that the antisymmetric spatial wavefunction of type (2) leads to a repulsive potential energy curve which they identified with the van der Waals repulsion between hydrogen atoms. In the same paper Heitler and London were the first to show that the ground state of the helium dimer is repulsive, and also formulated the general conclusion that those solutions of the Schrödinger equation that could represent binding between ground state rare gas atoms are forbidden in quantum mechanics.

The VB method itself has been found later to be perfectly suitable for advanced and accurate computations of molecular interaction energies and has been adapted for this purpose by several groups [44]. Similarly as in the perturbation theory approach the deformation and correlation effects can be accounted for by including wavefunctions corresponding to single excitations for the former and to double excitations for the latter effects, respectively. The convergence of the expansion can be improved by using for "excited" subsystems not the virtual SCF orbitals but those suitable for calculation of polarizabilities.

Similarly as the perturbation theory the valence bond method gives the interaction energy in terms of physically meaningful contributions, i.e., the electrostatic, induction and dispersion energies. It also offers the possibility to avoid the basis set superposition error that plagues all supermolecular MO type calculations.

5. CONCLUDING REMARKS

At the First International Congress of Quantum Chemistry in Menton E.R.Davidson started his lecture with the statement that "the fundamental goal of quantum chemistry is the development of a qualitatively correct set of concepts for describing the chemical and physical properties of molecules" [45]. Most of us agree with this statement and therefore the importance of the Heitler-London work and its significance for the development of chemistry are quite obvious. It was not only (in conjunction with the paper by Y.Sugiura) the first complete quantum mechanical *ab initio* treatment of a molecule and the first satisfactory explanation of its stability. The Heitler-London theory, extended by Slater and Pauling, provided rationalization of a variety of chemical phenomena. Let us recall once again the vast area covered by the VB treatment: covalent and ionic bonds, directional and energetic properties of bonds, multiple bonds, saturation of valence, specific properties of coordination ions, of conjugated or aromatic molecules, and the whole field of molecular interactions.

It is true that most of the modern *ab initio* approaches to molecular systems are of the MO type, and they provide reliable quantitative answers to chemical questions. Therefore one can notice that the chapters on the VB theory shrink in quantum chemistry textbooks, or disappear completely. This does not, however, change the fact that the concepts and the type of reasoning introduced in the VB approach are in everyday use among chemists who quite often even do not realize what is the origin of these concepts. Let us also emphasize, however, that today the VB approach is not a qualitative theory as it mostly was in the classical formulation. Generalized and adapted to the present day computing facilities it is successfully used to get accurate quantitative results for molecules and molecular complexes.

ACKNOWLEDGEMENTS

I am greatly indebted to the Organizing Committee of the Sixth International Congress of Quantum Chemistry in Jerusalem, and in particular to Professor Joshua Jortner, for making my participation in the Congress possible and for inviting me to present a plenary lecture. I am also grateful to B.Jeziorski for interesting discussions and to Professor J.P.Dahl for kindly providing me with valuable historical information.

BIBLIOGRAPHY

1. G.N.Lewis, J.Am.Chem.Soc. **38**, 762 (1916).
2. H.Hellmann, Einführung in die Quantenchemie, Deuticke, Leipzig, 1937.
3. E.Schrödinger, Ann.Phys. **79**, 361 (1926).
4. W.Heisenberg, Z.Physik **38**, 411 (1926); **39**, 499 (1926); **41**, 239 (1927).
5. W.Heitler and F.London, Z.Physik **44**, 455 (1927).
6. M.Born and R.Oppenheimer, Ann.Phys. **84**, 457 (1927).
7. E.U.Condon, Proc.Natl.Acad.Sci. USA **13**, 462 (1927).
8. F.Hund, Z.Physik **40**, 742 (1927).
9. Ø.Burrau, K.Danske Vidensk.Selsk. **7**, 1 (1927).
10. E.U.Condon, Proc.Natl.Acad.Sci. USA. **13**, 466 (1927).
11. E.U.Condon, Verh.d.Deutsch.Phys.Ges. **8**, 19 (1927).
12. A.Unsold, Ann.Phys. **82**, 355 (1927).
13. Y.Sugiura, Z.Physik **45**,484 (1927).
14. S.C.Wang, Phys.Rev. **31**, 579 (1928).
15. N.Rosen, Phys.Rev. **38**, 2099 (1931).
16. S.Weinbaum, J.Chem.Phys. **1**, 317 (1933).
17. C.A.Coulson and I.Fischer, Phil.Mag. **40**, 386 (1949).
18. L.Pauling, Proc.Natl.Acad.Sci. USA **14**, 359 (1928).
19. J.C.Slater, Phys.Rev. **37**, 481 (1931); **38**, 1109 (1931).
20. L.Pauling, J.Am.Chem.Soc. **53**, 1367 (1931); **54**, 988 (1931).
21. L.Pauling, The Nature of the Chemical Bond, Cornell University Press, Ithaca, New York, 1960.
22. W.Heitler, in Marx Handb. d. Radiologie II, 485 (1934).
23. R.Eisenschitz and F.London, Z.Physik **60**, 491 (1930).
24. F.London, Z.Physik **63**, 245 (1930); Z.Physik.Chem. (B) **11**, 222 (1930).
25. J.Gerratt, in Theoretical Chemistry, Vol.4, Specialist Periodical Reports, Chemical Society, London, 1974.
26. E.P.Wigner, Group Theory, Academic Press, New York, 1959.

27. R. Pauncz, Spin Eigenfunctions, Plenum, New York, 1979.
28. M. Kotani, A. Amemiya, E. Ishiguro and T. Kimura, Table of Molecular Integrals, Maruzen Co., Tokyo, 1955.
29. G. Rumer, Göttingen Nachr. 377 (1932).
30. H. J. Kopineck, Z. Naturforsch. **7a**, 314 (1952).
31. A. C. Hurley, Proc. Roy. Soc. (London) **A261**, 237 (1961).
32. W. Kołos and L. Wolniewicz, J. Chem. Phys. **45**, 409 (1966).
33. P. O. Löwdin, J. Chem. Phys. **18**, 365 (1950).
34. J. C. Slater, J. Chem. Phys. **19**, 220 (1951).
35. J. C. Slater, Quantum Theory of Molecules and Solids I, McGraw-Hill, New York, 1963, p. 74.
36. P. O. Löwdin, Phys. Rev. **97**, 1474 (1955); H. F. King, R. E. Stantom, H. Kim, R. E. Wyatt and R. G. Parr, J. Chem. Phys. **47**, 1936 (1967); F. Prosser and S. Hagstrom, Int. J. Quantum Chem. **1**, 88 (1967); **2**, 89 (1968); J. Chem. Phys. **48**, 4807 (1968); G. A. Gallup, Int. J. Quantum Chem. **6**, 899 (1972); G. G. Balint-Kurti and M. Karplus, in Orbital Theories of Molecules and Solids, ed. N. H. March, Clarendon, Oxford, 1974; M. Raimondi, W. Campion and M. Karplus, Mol. Phys. **34**, 1483 (1977); G. A. Gallup, R. L. Vance, J. R. Collins and J. M. Norbeck, Adv. Quant. Chem. **16**, 229 (1982); S. C. Leasure and G. G. Balint-Kurti, Phys. Rev. **A31**, 2107 (1985).
37. J. C. Slater, Phys. Rev. **35**, 509 (1930).
38. See, e.g., W. A. Goddard, T. H. Dunning, W. J. Hunt and P. J. Hay, Acc. Chem. Res. **6**, 368 (1973); W. A. Goddard and L. B. Harding, Ann. Rev. Phys. Chem. **29**, 363 (1978).
39. See, e.g., J. Gerratt and M. Raimondi, Proc. Roy. Soc. (London) **A371**, 525 (1980).
40. D. L. Cooper, J. Gerratt and M. Raimondi, in Ab Initio Methods in Quantum Chemistry, Vol II, ed. K. P. Lawley, John Willey & Sons Ltd., New York, 1987.
41. A. C. Hurley, J. Lennard-Jones and J. A. Pople, Proc. Roy. Soc. (London) **A220**, 446 (1953).
42. C. R. Mueller and H. Eyring, J. Chem. Phys. **19**, 1495 (1951); C. R. Mueller, J. Chem. Phys. **19**, 1498 (1951).
43. See, e.g., B. Jeziorski and W. Kołos, in Molecular Interactions, Vol. III, ed. H. Ratajczak and W. J. Orville-Thomas, John Willey & Sons, Chichester, 1982; W. Kołos, in New Horizons of Quantum Chemistry, ed. P. O. Löwdin and B. Pullman, D. Reidel, Dordrecht, 1983.
44. See, e.g., P. E. S. Wormer, T. van Berkel and A. van der Avoird, Mol. Phys. **29**, 1181 (1975); P. E. S. Wormer and A. van der Avoird, J. Chem. Phys. **62**, 3326 (1975); P. Cremaschi, G. Morosi, M. Raimondi and M. Simonetta, Mol. Phys. **38**, 1555 (1979).
45. E. R. Davidson, in The World of Quantum Chemistry, ed. R. Daudel and B. Pullman, D. Reidel, Dordrecht, 1974.

INDEX